Education Division 2018

Core Programming Area at the 2018 AIChE Annual Meeting

Pittsburgh, Pennsylvania, USA
28 October - 2 November 2018

ISBN: 978-1-5108-7629-3

Printed from e-media with permission by:

Curran Associates, Inc.
57 Morehouse Lane
Red Hook, NY 12571

Some format issues inherent in the e-media version may also appear in this print version.

Printed by Curran Associates, Inc. (2019)

For permission requests, please contact AIChE
at the address below.

AIChE
120 Wall Street, FL 23
New York, NY 10005-4020

Phone: (800) 242-4363
Fax: (203) 775-5177

www.aiche.org

Additional copies of this publication are available from:

Curran Associates, Inc.
57 Morehouse Lane
Red Hook, NY 12571 USA
Phone: 845-758-0400
Fax: 845-758-2633
Email: curran@proceedings.com
Web: www.proceedings.com

TABLE OF CONTENTS

Some abstracts in this book refer to images that were not made available by the publisher

541621 Presentation and Workshop on Career Planning

Abstract:

The transition to a faculty position can be a difficult process. This workshop is intended to help prepare prospective faculty prepare for the search process and successfully transition into an academic position. It will offer advice on identifying suitable career goals and specific objectives, and developing a plan to achieve them. Special attention will be devoted to establishing a funded research program. Elements of the career plan will include:

- Navigating the Job Search Process (research and teaching plans, interviewing, start-up packages)
- Establishing a Research Program (obtaining funding, Career Award preparation, problem identification & selection, contacting funding sources, network development)
- Understanding the University Environment (tenure & promotion, service activities, sabbaticals, reward structure, mentoring expectations)
- Managing Your Time (goal setting & priorities, stress reduction, good habits)

542555 Important Hard and Soft Skills Difficult to Teach in the Classroom

Abstract:

There are many important skills and experiences imparted through working on international humanitarian engineering projects. Understanding your customer, dealing with other cultures, safety, dealing with uncertainty, and thinking on your feet are a few of them. Students can benefit greatly from taking on this kind of work either in implementing full projects or doing the conceptual work in a capstone design class.

542581 A Student Perspective: What I Learned from Working on an International Project

Abstract:

Students volunteer to work on international projects for many different reasons. They take on many different roles in planning, design, execution, communications, project management, and research; and they each develop different skills, have different experiences and are impacted differently. Undergraduate students will discuss why *they* volunteered, what *they* have done, and what *they* have gotten out of this work professionally and personally.

542554 What Does Sustainability Really Mean in These Projects?

Abstract:

To make a project successful partnering with a small community in the developing world requires many important attributes besides technical accuracy. In order to sustain the benefits over the long term, the project team must consider and master many other aspects of sustainability in addition to environmental sustainability. Financial and cultural are two of the many types of sustainability that will be discussed.

542556 What Is EWB-USA and Why Might You Want to Work with Them?

Abstract:

EWB-USA is organization of volunteer students and professionals. Student and professional chapters can be found on campuses and cities across the country. EWB-USA has very specific standards in terms of how partnerships with local communities are maintained and projects executed. The EWB-USA philosophy is tightly focused on executing projects that will actually deliver value over the long-haul, and the mandatory work processes take into account the unique characteristics of student volunteers.

542585 How International Projects Might Help with ABET Accreditation

Abstract:

Team humanitarian engineering projects align with many ABET Criterion 3 Student Outcomes. Incorporating the projects into a programâ€™s assessment structure is difficult when the projects are extra-curricular. We will present solutions from different schools that allow the projects to assist accreditation.

542580 Panel Discussion -- Making Successful International Projects Happen

Abstract:

There are many challenges to implementing humanitarian engineering projects in a university setting. Where do the ideas / projects come from? How can these projects be turned into meaningful capstone design classes? How can they be fit into existing curriculum and academic environment? How do you motivate students, faculty, administration? How can faculty motivated to make these projects happen collaborate across different universities? This panel is intended to be a forum for all those in the room to define challenges and offer solutions.

525999 Fluidization Experiments in a Unit Operations Laboratory

Abstract:

Unit Operations Laboratory courses are ubiquitous in chemical engineering departments and often represent a rite of passage for upperclassmen. Key goals of the laboratory courses include reduction of theory to practice, technical report writing, and pilot scale equipment operation. Ancillary goals include enhancing soft skills with respect to group dynamics and opportunities to improve techniques related to technical presentations. Typically, the Unit Operations Laboratory is taught in the junior and senior years, when most of the core courses have been completed. The theoretical framework for analyzing relevant chemical engineering problems has been established, and students should be able to analyze data obtained in a laboratory setting relative to the theory. Some of the most important aspects of the laboratory experience involve determination of the reasons why the data from an experiment does not match the theory. Students learn practical equipment and instrumentation handling skills and laboratory safety best practices.

The Unit Operations Laboratory at The Ohio State University (OSU) includes two different experiments focused on fluidization phenomena. Fluidization is a technology that is often omitted from an undergraduate curriculum or given a high level overview at best. While this technology is not often covered in depth, there are numerous industrially relevant examples of fluidization, and students benefit from the practical aspects of the experiments. This submission will summarize the two fluidization experiments in the Unit Operations Laboratory at OSU. This will include the relevant theory and applications for fluidization and details on the two experiments. The physical construction of the experiments and data capture for the experiments are relatively simple and inexpensive and make these experiments candidates for other departments looking to diversify their Unit Operations experiments.

In the gas-solid fluidization experiment, students investigate fluidization phenomena including the pressure drop and bed height as air flows through a packed bed. The Ergun equation is used to model behavior in the column, and students compare the minimum fluidization velocity predicted by the Ergun equation with that measured in the lab. Differential pressure meters are used to quantify the pressure as a function of position in the column. Glass spheres with diameters ranging from 300-800 microns are used in this experiment to allow students to measure how the performance of a fluidized bed is impacted by particle size and particle size distribution. Some members of the group on this experiment perform porosity measurements, using both wet and dry methods, with the data used for subsequent calculations of minimum fluidization velocity.

A gas-solid-liquid fluidization column complements the gas-solid fluidization experiment. Students learn about different flow regimes in three phase fluidization and explore the details of wake theory in the context of mass transfer in a fluidized bed. The main column has a diameter of 0.19 meters and a height of over five meters. Glass beads are loaded in the bottom of the column, water fills the column, and an air stream flows upward through the column. There are fifteen pressure ports at various locations on the column, with a manometer bank near the top of the column to monitor dynamic pressures. A smaller diameter recycle line runs parallel to the main column, and a valve controls the flow through the recycle line, permitting experiments to be performed with and without water recycle. Similar to the gas-solid fluidization experiment, students investigate the impact of superficial velocity on bed performance.

Both fluidization columns are fabricated from acrylic sections that are adhesively welded together or joined with gasketed flanges. Either apparatus could be constructed for a total cost of less than $5,000, thus making either experiment viable for a Unit Operations Laboratory.

537561 Propagation of Hands-on Desktop Learning Pedagogy across Institution and Program Types

Abstract:

Low Cost Desktop Learning Modules (LC-DLMs) were created at Washington State University (WSU) within the Chemical Engineering program to aid in student comprehension of a variety of engineering concepts. The LC-DLMs are hands-on apparatuses that can be used to represent the theories behind process units seen in industry. Activities associated with the modules may be used to supplement lecture material. Although the LC-DLMs have been proven to work within the initial classroom setting, the pedagogy will not reach its full potential without translating to other disciplines and outside institutions. Because of demonstration through the initial years of testing at WSU, the engineering education community has gained awareness of the LC-DLMs, but the next step requires propagation of this learning method. Although the terms "propagation" and "dissemination" are commonly used interchangeably, they act as two separate steps that lead to the overarching goal of adoption; dissemination is the sharing of methods, while propagation is successful use of such methods by professors in other disciplines and at outside institutions. Teaching seminars, online platforms, and instructional videos are a few methods considered to ease the transition of a new implementation into the course curricula at beta institutions. To test the translatability of the LC-DLMs a set of modules were implemented within chemical engineering and/or mechanical engineering at WSU, the University of Kentucky – Lexington and Paducah, University of Central Oklahoma, and University of Idaho. In this study, student assessment results from the beta implementations will be presented, as well as instructor feedback on ease of use.

537980 Design of an Automated Control System for a Continuous Distillation Column and Its Implementation into the Unit Operations Laboratory Experimental Program

Abstract:

A research program is described that transforms a manually operated continuous-flow, pilot-scale distillation column into a fully automated system based upon a process logic controller (PLC) with a customized operator interface. The new system provides an advanced platform to investigate both steady-state and process dynamics of either binary or multicomponent distillation to support both student training and research in advanced separation processes. The automation package is based upon a Siemens series 505 PLC with a Wonderware Intouchâ„¢ human machine interface (HMI). Various field instruments used to measure local pressures, temperatures, flow rates and specie concentrations are connected to the PLC using a special-purpose hardware interface. The process and instrumentation diagram, instrumentation, operator interface and experimental protocols are described along with process safety aspects. The HMI is configured so the user can operate the system using either manual or automatic modes. The user can define a timed-sequence for the process input variables from which the corresponding system output responses can be displayed in real-time. An advanced experimental program developed for the undergraduate level Unit Operations Laboratory courses based on the automated system is described. This experimental program will enhance student educational outcomes in the core Chemical Engineering curriculum areas of separations, mass and heat transfer, phase equilibria, and process control.

529470 Process Control Laboratory on Arduino and Simulink Platform

Abstract:

Process Control Laboratory on Arduino and Simulink Platform

Yuanyi Wu [1], Sohrab Rohani [1*]

1. The University of Western Ontario, Department of Chemical and Biochemical Engineering, London, Ontario N6A 5B9, Canada

* Corresponding authors: srohani@uwo.ca

Process control and control theory involve both abstract mathematical concepts and concrete controller implementation. In process control education, the hands-on lab plays a significant role in building connection between the theory and the application and helping students gain experience in engineering practice. We proposed a low-cost process control experiment platform based on the Arduino board (Figure 1) and MATLAB Simulink. Students were inspired to build an experiment setup from scratch by themselves and learn the essential elements in the control system. In this teaching case, Simulink provided a codeless block-diagram based modelling environment with handy process identification, parameter estimation, and controller design tools. The Arduino microcontroller codes can be generated from the block diagram model by Simulink Arduino hardware support package (Figure 2). Without learning a new programming language, students can utilize the symbols and the block diagram representations that are consistent with the textbook to implement and validate the controller performance taught in class. The lab session progressively introduced the topics including handling the analog inputs and outputs, processing the data with filters, estimating the dynamic model with collected data, and tuning a PID controller with the model. The students' feedbacks indicated that the hands-on lab helped to understand the control theories and motivated them to employ the modelling and rational controller design methods in their research projects.

Figure 1 Arduino temperature control lab platform

Figure 2 Simulink block diagram can run on the Arduino board with real-time data visualization and controller parameter tuning features   Acknowledge

The authors acknowledge the financial support provided by the Natural Science and Engineering Research Council (NSERC) of Canada.

532082 Experimental System and Process Data Describing Model-Based Control Strategies in the Chemical Engineering Curriculum

Abstract:

The two pillars of control system design in chemical process engineering applications involve either feedback or model-based approaches like Internal Model Control (IMC). Feedback control has been the industrial standard for decades and while it is robust and straightforward to implement has the drawback of not being agile enough to deal with processes exhibiting long time delays. These delays are significant in many chemical processes. However, the combination of model-based methods, with the Smith predictor concept, provides an excellent alternative for systems with long time delays. We will discuss a novel experimental system designed to illustrate the advantages of this latter approach for a system showing significant time delays which we have included as an experimental component to our senior year process control course in the chemical engineering department. The system has served as an invaluable tool in enhancing our students' understanding of theoretical material taught in lectures and should be of interest to the larger chemical engineering educational community.

538750 Creative Student Activities to Enhance Teaching on Heat and Mass Transport

Abstract:

Transport Phenomena and Heat and Mass Transport (HMT) in particular are courses that have been taught to generations of Chemical Engineers. Such courses typically form part of the curriculum for Junior Chemical Engineering students. Popular textbooks for Heat and Mass Transport are often latest editions of older, quite successful texts, while homework problems that are given to students are very similar to the homework problems that the instructors solved themselves when taking these courses several years ago. How can the students be more engaged with the material covered in HMT? In the form of extra-credit assignments, Juniors in Chemical Engineering at the University of Oklahoma were asked to exercise their creativity and utilize their talents to generate original creative product that is related to the class material. The students were free to develop a product that could be a video, a graphic novel, a comic story, a music synthesis etc. but related to HMT. More than 80% of 122 students chose to work on such a project. It turns out that each student put on average more time in preparing their creative project than they spent in preparing for one of the tests for the class. In terms of assessment, an anonymous student survey showed that students who engaged in preparing a creative project for HMT felt that they understood the material better as a result of engaging in this activity. Demonstrations of student products will be presented, as well as assessment results for the effectiveness of this approach in enhancing student learning and increasing interest in the course.

533856 Chemical Engineering 'on-a-Chip': Capturing the Integrated Scope of Chemical Engineering in a Single STEM Module

Abstract:

Unlike other engineering disciplines, such as electrical or mechanical engineering, where the scope of study and career paths are clear, the skill sets, impact and contributions of chemical engineers (ChEs) and the broad spectrum of jobs that they hold tend to be poorly understood. Specifically, there exists a persistent struggle to differentiate ChEs from chemists, and a failure to grasp how ChEs leverage all of the basic sciences for designing processes capable of transforming matter into new valuable products. This can present challenges for recruiting students into the study of chemical engineering. Yet, the same complexity that makes the chemical engineering discipline and its impact difficult to grasp also challenges educational initiatives aimed at conveying, in a single module, a sufficient representation of its scope. While chemical engineers often think about challenges associated with process scale-up, demonstration of large-scale process integration within the confines of the STEM laboratory can be logistically challenging and costly. As a viable alternative, the involvement of chemical engineers in process scale-down—for example, in the design of molecular sensors, miniaturized devices for health assays and drug discovery, or chemical plants on-a-chip—provides an attractive basis for developing cost-effective, yet impactful ChE-based STEM modules.

In this talk, we will describe a hands-on educational module created to illustrate the full scope of chemical engineering.[1] Students are taught fundamental concepts that they integrate in a laboratory experience involving designing and fabricating a 'plant-on-a-chip' microfluidic device capable of continuous processing of reactive flows. Students gain insight into ChE principles in the context of process scale-down, including extracting and extending reaction kinetics from bench to microfluidic-scale while accounting for the need for enhanced mixing within laminar fluid flow. This integration of ChE fundamentals offers insight into chemical engineering process design principles that are commonly challenging to capture in single STEM modules. The active nature of this learning module teaches students about real challenges faced by chemical engineers in their work environment, effectively educating students about the vastness and impact of the discipline and aiding in the recruitment of students into the discipline.

References

1. K.M. Schultz and M.A. Snyder, "Chemical Engineering 'On-a-Chip': Capturing the integrated scope of chemical engineering in STEM outreach," *invited article*, under revision, *Chemical Engineering Education (CEE)* (2018).

538778 The Use of Numerical Worksheets in Undergraduate Courses

Abstract:

Experience in using a user-friendly software, Mathcad, in the material and energy balance course and chemical reaction engineering course is discussed. Example problems considered for illustration deal with solution of linear algebraic equations (parameter estimation, mass balances, independent reactions), nonlinear algebraic equations (equilibrium calculations for multiple reactions, vapor-liquid equilibrium calculations for mixtures, solutions of equations of state, design equations for steady state well-mixed reactors), ordinary differential equations (isothermal, adiabatic and non-adiabatic batch and steady state continuous flow tubular reactors), and symbolic algebra (rate expressions for catalytic reactions). The capabilities of Mathcad are of significant benefit in accelerating the learning and strengthening the fundamental knowledge base.

530820 Active Learning in a Large Lecture Setting Enhances Introductory Course

Abstract:

At UC Berkeley we require first year and entering transfer students who have declared chemical engineering as their major to take a 2-unit course, "Introduction to Chemical Engineering Design." Enrollment in the course has been up to 150. The theme of the course is, "Chemical engineers design and analyze processes in which physical and chemical transformations yield useful products." I will present how I develop that theme through weekly lectures, discussion sections, and homework assignments and how I incorporate active learning in the classroom, asking the students to engage in each lecture through conversations with each other (think-pair-share activities) and brainstorming activities. The course culminates with a final team project related to energy and sustainability where students prepare written memos and deliver oral presentations; past topics have included: recycling lithium ion batteries, thermal salt storage, desalination, carbon capture and conversion to methanol, and converting food waste to energy. Some of the non-technical outcomes of this course include leveling the knowledge base of the students before they proceed in the curriculum and fostering an attitude of cooperation vs. competition.

539675 Peer Led Team Learning in Chemical Engineering

Abstract:

Peer Led Team Learning (PLTL) is an evidence-based model of teaching undergraduate courses in STEM fields. It has been shown to provide benefit both to students and peer leaders (Wilson and Varma-Nelson). Students saw statistically higher course grades and self-reported better understanding of material and improved logic. Peer leaders self-reported better communication and facilitation skills and better understanding of content. Several PLTL studies have also suggested increased problem-solving and critical thinking skills of participating students (Wilson and Varma-Nelson). The bulk of the literature data is based upon studies in chemistry, physics, and mathematics courses. Best practices from several studies were combined to pilot a PLTL program in an introductory Chemical Engineering Computations (Matlab) course. Here, we will explore the best practices that ultimately led to a 0.2 GPA improvement (on a 4.0 scale) and improved course evaluations across all categories of assessment.

Wilson, S.B. and P. Varma-Nelson. "Small Groups, Significant Impact: A Review of Peer-Led Team Learning Research with Implications for STEM Education Researchers and Faculty." *Journal of Chemical Education* 93 (2016): 1686-1702.

518366 Teaching Students How to Learn: Blending Retrieval Practice with Team-Based Learning in a Material and Energy Balances Course

Abstract:

This presentation will describe how two high-impact, evidenced-based instructional strategies, Retrieval Practice and Team-Based Learning, were implemented to update the course delivery format used in a material and energy balances course for second-year chemical engineering students. The talk will provide examples to illustrate how these changes in course delivery impact what students learn and how students go about learning it.

530301 A New Curriculum to Train Chemical Engineers to Solve 21st Century Grand Challenges

Abstract:

The Department of Chemical and Biological Engineering at the University of Sheffield is embarking on a curriculum change project with roll out starting with level 1 in September 2017. The drivers behind the change include the need to modernise the curriculum both in terms of content, structure and delivery. The main objective is to develop a modern Sheffield Chemical Engineer. The study is primarily about investigating the efficacy of the change efforts that have been introduced, to track progress and to determine whether we are meeting our stated objectives. The objectives are in relation to student success, student experience, curriculum coherence and student and staff well-being. Specifically, the new curriculum will be coherent, embedded in design and practice with an emphasis on critical thinking, problem solving, professionalism, ethics and sustainability. It will offer flexible learning environments and pathways to facilitate deep engagement. It will promote and facilitate industry involvement by focusing on both process and product engineering to develop industry ready practical graduates with hands on experience. It will produce graduates who are integrators, change agents and self-directed learners to lead multidisciplinary teams, and be at the forefront of innovation. It will provide exposure to niche research areas built on a strong core in engineering fundamentals. Lastly, it will produce graduates capable of Engineering from molecules by applying systems level thinking at many length scales. We have identified a third year module process design as a significant check point to determine whether some of our curriculum objectives are being met.

519808 Using Online Survey Tools to Facilitate Classroom Discussion on Process Safety

Abstract:

Process safety and ethics within Chemical Engineering continues to be a strong topic of focus. Students are continually challenged to be mindful of soft skill issues associated with being a professional engineer; however, the retention of that knowledge is low. The prevalence of increasing student awareness in process safety and ethics beyond the textbook requires innovative ways to bridge the discussion between fundamental course content and experiential learning. Given the time constraints of contact hours an instructor may have for a course, the integration of practical discussions can be pushed aside to ensure the learning outcomes are met. This, however, can be a disservice to the students as they prepare to take up jobs either in a co-op/internship or full-time position. One approach to overcoming the time constraints associated with limited contact hours is to seed the conversation of process safety and ethics through the use of scenario-based surveys. Using applications, like SurveyMonkey, can help to facilitate discussion on different topics associated with process safety and ethics. Providing real-time examples allows for stimulating discussions when time is limited. A discussion on the use of ready-made technology to facilitate process safety and ethics in the classroom will be shared. A primary focus will be on implementing the approach while not sacrificing classroom contact time.

524605 Introducing Chemical Engineering to Prospective Engineers

Abstract:

A significant challenge often faced for students considering chemical engineering as a major is attempting to grasp the idea of what exactly chemical engineers do and what chemical engineering is about. However, an issue that is often less addressed is the significant lack of understanding students in other engineering majors have about chemical engineering, a larger issue when compared to other engineering fields. Because of the breadth of roles chemical engineers take both in an academic setting and in industry, they will often work with a wide variety of engineers with different expertise, meaning that it would be of great interest for them to gain at least a basic idea of what chemical engineering is about. In an attempt to bridge these gaps in understanding, a half-semester module was developed in an Introduction to Engineering class at Lafayette College. Early offerings of the module have been generally well-received by students in all majors, reporting a better grasp of what chemical engineers do (even if they have absolutely no desire to major in it). This talk will describe the experiences, input and constraints that led to the development of the module, a discussion of the implementation and subsequent modification of the module and feedback from the students along with self-reflection.

530979 Ten Years in the Trenches: An Updated Suite of Scenario-Based Academic Integrity Videos

Abstract:

Academic integrity in the classroom has always been a concern, but the methods and opportunities for students to take shortcuts have increased dramatically in recent years with advances in technology. In the past faculty members worried about studentsâ€™ copying each otherâ€™s homework or looking over someone elseâ€™s shoulder during an exam. However, the introduction of solution sharing websites and smart phones has provided students with numerous opportunities to get unauthorized aid on homework, exams, projects, and lab reports. In response, the authors developed a suite of scenario-based academic integrity videos ten years ago which were effective for a time, but have become outdated in recent years. Moreover, the scenario-based videos were limited in scope to a single course in a single engineering discipline, which restricted their overall utility across the curriculum. This presentation will outline the design and show examples of a new suite of scenario-based videos updated to address new challenges facing course instructors. The videos feature six segments consisting of eight to twelve separate scenes dealing with academic integrity issues using online resources, peer resources, instructor resources, computer problems, reports/projects, and exams. The authors worked closely with a group of undergraduate students to develop the content and utilized these students as both writers for the script and actors in the video. All of the scenarios were modified to reference examples in calculus, chemistry, and physics courses to make them more widely applicable across a broader range of science and engineering disciplines. The videos are all open source and can be accessed through an unsearchable YouTube page to allow dissemination to any instructor who chooses to utilize them. We will also offer suggestions on how to utilize the videos along with additional academic integrity-related resources, such as syllabus language, a reflection assignment, and an assignment cover sheet to supplement the message in the videos.

522682 Evaluation of Conceptual Testing Enhanced with Technical Writing and Just-in-Time Teaching

Abstract:

This study investigates the effectiveness of a combination of just-in-time teaching and conceptual testing (â€œconcept quizzesâ€) requiring either written or multiple-choice answers toward improving student performance in an introductory material and energy balances course. Multiple linear regression models, pairwise hypothesis testing and ANOVA analyses were used to determine if treatment groups affected student performance compared to a control group. It was found that a combination of written-answer concept quizzes and just-in-time teaching activities significantly increased (p=0.003) student scores on an AIChE Concept Warehouse testing inventory. However, no tested treatment had a significant effect on student exam performance, even on exam questions related to concepts stressed in concept quizzes; it is suggested this finding is due to tested exam questions having a computational rather than conceptual focus. A lingering concern for instructors assigning written-answer concept quizzes is that ESL students may be disadvantaged due to additional language translation steps which are not faced by native-speaking students. This concern was investigated using a nonparametric Wilcoxon-Mann-Whitney test which indicated that ESL status of students assigned written-answer concept quizzes was not found to have a significant effect on various measures of student performance (p=0.47-0.77).

527523 Prediction of Optimal Chemotherapy Dosing Regimens: Balancing Tumor Degradation and Toxicity Effects

Abstract:

During chemotherapy, clinicians must constantly balance the dose of chemotherapeutic agents to reduce the tumor volume and minimize toxicity to the patient. The theory of optimal control can be applied for better assessment of growth-toxicity dynamics and development of drug dosage schedules. The reduction of circulating leukocytes or white blood cells (WBC) by chemotherapeutic agents (myelosuppression) is of primary concern and is often the dose-limiting factor in treatment. For these reasons, the effect of chemotherapy on a patientâ€™s WBC population is considered as a metric of toxicity in the development of optimal dosing regimens. A model is developed to include pharmacokinetic compartments to simulate the concentration of the anticancer agent in distinct parts of the body. Additionally, a more complex model of myelosuppression with demonstrated accuracy is incorporated to analyze the immunotoxic effects. This combined model is evaluated via discretized non-linear programming methods to gain insight into optimal dosage regimens for a specific anticancer agent. This model will aid in quantitating hematological toxicities and will contribute to the design of reliable, personalized tools to optimize chemotherapy and improve patient outcomes.

540910 Effects of Waste Vegetable Oil in High Concentration Recycled Asphalt Pavement Binder Mixes

Abstract:

As the worldâ€™s focus has shifted toward environmental sustainability and the costs of roadwork have put economic stress on governments, researchers have been investigating viable ways to increase the use of Recycled Asphalt Pavement (RAP). Simultaneously, there is critical importance to find use for large quantities of waste cooking oil produced annually in the United States. The objective of this research is to investigate the rejuvenating effects of Waste Vegetable Oil (WVO) in high RAP binder concentration mixes, as well as the effects these mixes experience when aged. The ability to utilize WVO in high concentration RAP mixes would be incredibly beneficial because of the much lower costs of road materials, lower amounts of asphalt and vegetable oil waste, and the possibility of road conditions deteriorating less rapidly. The RAP binder used in this experiment was simulated using accelerated aging methods with a Rolling Thin Film Oven (RTFO) and Pressure Aging Vessel (PAV). A single type of vegetable oil (Mel-Fry Essential) was used to isolate the effects of chemical composition. Different ages of the oil during food preparation were collected to measure the changes in properties with use. Binder mixes were prepared with differing concentrations of oil (0 and 10%), RAP (0, 40, and 60%), and virgin binder. These mixesâ€™ rheological properties were measured when unaged, RTFO aged, and PAV aged. The rheological properties of these RAP mixes were tested through the means of Rotational Viscometry (RV) and Dynamic Shear Rheometry (DSR). The results of these experiments have shown that the WVO used can effectively lower the viscosity and stiffness of high concentration RAP binder mixes. Further, WVO in RAP mixes has shown to provide significant resistance to effects of aging over time. Because of these findings, future research on WVO in binder mixes will be conducted to investigate the mechanism of this aging resistance, as well as the performance of the binder and mixes in other tests such as Bending Beam Rheometry (BBR), Dynamic Contact Angle (DCA), and moisture resistance.

529127 Microfluidic Modulation of Neural Differentiation of 3D Stem Cell Aggregates

Abstract:

Embryonic stem cells (ESCs) can be differentiated as aggregates into specific cell and tissue types for a range of applications. However, current studies are limited by culture methodsâ€"in particular, the ability to control and perturb the culture environment and to image and assess samples throughout culture. To address these challenges, we developed a microfluidic platform that provides a controlled environment for culture and differentiation of mouse embryonic stem cell (mESC) aggregates. This platform allows us to image samples *in situ*, track samples longitudinally, and control the rate of media exchange. Microfluidic media exchange can be used to alter concentrations of cell-secreted factors in the local culture environment. Our goal was to assess how the residence time of media components and cell-secreted factors affect cell growth and differentiation to neural cell types.

Results showed that continuous media exchange was detrimental to generation of progenitor motor neurons, potentially because necessary cell-secreted factors were continuously removed. In contrast, exchanging media at defined intervals increased generation of progenitor motor neurons in microfluidic cultures. Interestingly, we also observed asymmetric patterns of differentiation in a subset of cell aggregates: portions of aggregates better protected from flow were observed to have increased numbers of motor neuron progenitors. Together, our results provide insight into how microfluidic media exchange can modulate cell behavior and differentiation. This can inform design of microfluidic culture conditions in future studies of directed differentiation and tissue morphogenesis.

531268 Building Better Proteins: Integrating Cell-Free Protein Synthesis and Coarse-Grained Molecular Simulation to Rapidly Determine the Optimal Location for PEGylation

Abstract:

Protein therapeutics are life-saving drugs that combat some of the worldâ€™s most debilitating and painful diseases including cancer, multiple sclerosis, rheumatoid arthritis, Hepatitis, and diabetes. Many protein drugs are currently under development, and within five years, the number of protein therapeutics prescribed is expected to eclipse the number of small molecule therapeutics. This is particularly exciting as there is great hope for effective protein therapeutics to combat Alzheimerâ€™s, cancer, and genetic disorders. However, fundamental limitations of protein therapeutics include susceptibility to degradation and aggregation, stimulation of an immune response, and rapid clearance from the body. These impediments have led to the development of second generation drugs in which a protein therapeutic is covalently conjugated to the FDA-approved polymer polyethylene glycol (PEG). This has resulted in significant improvement in overcoming the above limitations; however, PEGylation typically leads to reduced protein activity. To overcome this drawback, our lab has developed a method to 1) site-specifically attach PEG to proteins and 2) find the optimal conjugation location that stabilizes the protein while maintaining its activity. The method utilizes a hybrid computational/experimental approach to determine the locational impact of PEGylation on PEG attachment efficiency, protein stability and activity. This is done by integrating a coarse-grain simulation approach with cell-free protein synthesis experimental technology. Through experiments utilizing T4 Lysozyme as a model protein, we demonstrate the utility of this integrated approach and evaluate the predictive impact of currently held design guidelines for PEGylation including site solvent accessibility, conjugation site structure, PEG size, and double PEGylation. Our findings revealed that PEGylation efficiency, protein stability, and activity were not well predicted by the current guidelines, illustrating the need for a rapid screening approach such as the one reported herein. The results of this work have been recently reported in ACS Synthetic Biology: 7(2):510-521.

537528 Anionic Microgels for the Trypsin-Mediated Retention and Release of Therapeutic Proteins

Abstract:

Protein therapeutics are currently administered by intravenous, subcutaneous, or intramuscular injection. While injection provides an effective administration route, the discomfort it entails causes low rates of patient compliance. Thus, delivering protein therapeutics orally is more desirable and will promote patient compliance. In order to achieve oral protein delivery, the polymeric carrier needs to be able to withstand harsh gastric conditions to protect the fragile protein cargo, and have a targeted release mechanism at the site of the small intestine.

In this study, several polymer systems were designed for the oral delivery of a high isoelectric point protein therapeutic in response to binding free trypsin that is abundantly present in the small intestine.

Previous studies had shown that a displacement mechanism using complementary ionic interactions between the microparticles and model therapeutic proteins in solution was viable. A displacement mechanism for protein delivery was verified by loading the hydrogels with cytochrome c, a model high isoelectric point therapeutic, before incubating the loaded hydrogels with trypsin, the target protein. The amount of cytochrome c released and trypsin adsorbed were quantified. Results showed higher binding capacity in low ionic strength solutions than high ionic strength solutions, indicating that the release mechanism was predominated by ionic interactions, and not by specific affinity for the trypsin molecule.

To increase our systemâ€™s affinity for trypsin, the base polymer was modified with various hydrophobic monomers. Additionally, the base particles were surface modified with specific peptide ligands.

A base polymer with methacrylic acid, N,N-methylene bisacrylamide, and acrylamide was modified with hydrophobic monomers or peptides to increase affinity for the trypsin molecule and efficiently release cytochrome c in response to the presence of trypsin. Multiple formulations of hydrogels with various hydrophobic monomers in different proportions (i.e. 5, 10, 20, and 40 mol%) were synthesized using free radical polymerization. The successful incorporation of hydrophobic monomers into the microparticles was verified through Fourier Transform Infrared Spectroscopy (FTIR) by comparing characteristic amide, methacrylate, and aromatic peaks.

The base particles were synthesized and also surface modified with peptide ligands. The peptide ligands were incorporated using EDC-NHS bifunctional linker between the carboxylic acid groups of the polymer and free primary amines on the peptide. The coupling of the peptide was verified by detecting primary amines via fluorescamine assay and microBCA colorimetric assay.

The binding capacities of these particles for model proteins were assessed and compared to particles with hydrophobic monomer. While particles incorporated with hydrophobic monomers decreased binding capacity of the particles for cytochrome c, modifying microparticles with oligopeptide ligands increased the specificity that the particles had for cytochrome c and trypsin.

This study meant to design polymer systems for the oral delivery of high isoelectric point protein therapeutics in response to the presence of trypsin. Two platforms, one with varying proportions of hydrophobic monomer, and one with peptide ligands modified onto the surface, were synthesized. The binding capacity of these platforms for a model therapeutic, cytochrome c, and for trypsin, were quantified. Particles surface modified with peptide ligands showed increased specificity for cytochrome c and trypsin when compared to the unmodified controls.

530899 Desktop Learning Modules (DLMs) and Their Effects on Student Progression through Bloom's Taxonomy for Fluid Mechanics Concepts

Abstract:

The use of alternative and complementary learning methods has been explored for the past several decades to aid in student comprehension and retention of engineering concepts. Hands-on learning methods, specifically, have gained notoriety due to their potential for being more applicable to engineering students, as the majority of these students tend to be kinesthetic learners. To support this mode of learning, low-cost Desktop Learning Modules (DLMs) were created by Dr. Bernard Van Wie's group at Washington State University. DLMs are hands-on apparatuses representing industrial process units; activities associated with the modules may be used to assist student learning of engineering concepts. To determine the effectiveness of these DLMs, two types of modules—Hydraulic Loss and Venturi—were tested on two Chemical Engineering undergraduate classes at Washington State University. To account for the differences in teaching style of the instructors and ensure equal opportunity for success, the two classes alternated between being treated as the control group and the variable group for each lecture unit. To begin the study, all students were given an articulated pre-test to determine their incoming knowledge of mass and energy balances with respect to the DLM of interest. From there, the control group continued with lecture as normal, while the variable group ran mini-experiments on the DLM and filled out a worksheet corresponding to the experiments. All students were then given the same posttest on mass and energy balances with respect to the DLM used. Both the pre- and posttests consist of a mix of multiple choice questions and justification sections that correspond to the Bloom's taxonomy levels: remember, understand, apply, analyze, evaluate, and create. This study will determine the effectiveness of the DLMs on student learning of fluid mechanics concepts using Bloom's taxonomy to observe student progression from pre- to posttest in the control and experimental groups.

540911 Characterization of Polymer Binders to Improve Cyclability of Lithium-Sulfur Batteries

Abstract:

New emerging technologies such as hybrid electric vehicles and fully electrified vehicles require higher energy densities than Li-ion can provide. As a result, we must look into different battery chemistries that can supply adequate power to these systems. Li-S has been identified as a potential candidate to replace the Li-ion. There are many issues preventing commercialization of the Li-S battery and it's poor cycling performance is a major obstacle to be overcome. This work focuses on using polymer binders to prevent the detrimental "shuttle effect" in these batteries. Poly-acrylic acid (PAA), carboxymethyl cellulose (CMC) and a tri-block copolymer (SBAS) were identified as potential candidates for use as binders in Li-S batteries. Various electrochemical techniques and SEM were used to characterize their abilities to improve cycle performance. From an analysis of the charge capacity and coulombic efficiencies vs the cycle number, a mixture composed of PAA:SBAS was shown to have the quickest and most stable cycling performance. Addition of SBAS to CMC or PAA improved the initial capacity and performance of the battery. SEM images revealed that the choice of binder could affect the sulfur distribution on the electrode. Electrochemical impedance spectroscopy (EIS) was used to study the resistances inherent with these binders. We proposed that the two-time constants observed in the spectra was similar to Gamry's proposed model of a failed coating. The first semi-circle was attributed to the capacitance across the polymer coating and the second semi-circle was attributed to the double layer formed near the electrode.

529257 Efficient Coacervate Extraction of Cationic Industrial Dye from Wastewater

Abstract:

Effluent wastewater containing dyes from textile, paint, and various other industries have long posed environmental issues. Functional nanomaterials offer new opportunities to treat these effluent wastes in an unprecedentedly high efficient and facile fashion due to their large surface area-to-volume ratio. In this work, we explore a new mechanically stable hybrid nanomaterial based on the complexation of zwitterionic polyampholyte with anionic polyoxometalate (POM) nanocluster to demonstrate efficient coacervate extraction of cationic dye, methylene blue (MB) as a model waste from its salted aqueous solution. MB wastewater samples were prepared at a concentration of 0.2 g/L and were used to subject the hybrid complexes to multiple rounds of extraction to determine the upper limit of adsorption of the material. Efficiency of coacervated complexes at varied POM concentrations, varied molar ratios of polyampholyte, and different salt concentrations were measured using a UV-vis spectrophotometer. Our polyionic complexes exhibited ~99% removal efficiency for all initial MB wastewater samples with measured adsorption capacity as high as 415 mg/g, indicating extremely high efficiency. Our study also suggested that coacervate extraction could be a facile and robust method for ionic selective wastewater treatment with parameters such as temperature and pH showing no influence on removal efficiency. With the simple and straightforward coavervate complexation in solution, coacervate nanomaterials advantageously possess dual processing options in separation industry through either membrane fabrication or use directly in mixer-settlers.

532559 Estimation of Critical Exponents Using Asymptotic Approximants

Abstract:

The virial equation of state is a power series in number density that incorporates attractive and repulsive interactions between molecules, which gives it the ability to be one of the best tools for modeling fluid behavior. Unfortunately, the virial series is expressed in terms of bulk fluid properties, so, although the virial series is theoretically infinite, coefficients are expensive to compute. Additionally, if there is a thermodynamic critical point, this sets the seriesâ€™ radius of convergence, which can make it difficult to predict critical properties using the virial series.

This project primarily analyzes the effectiveness of asymptotic approximants, which incorporate the virial series and known critical scaling laws. This method will be used to estimate the critical exponent along the critical isotherm. A case study is also conducted in estimating the critical exponent for the high temperature susceptibility expansion of the Ising model and comparing the estimates between the asymptotic approximant and more traditional methods. The aim of this work is to provide a methodology for constructing equations of state and predicting fluid properties over a broad range of fluids and conditions.

533596 Developing Engineering Thinking during an Ill-Structured Activity for Kinetics and Reactor Design: A Technology Solution

Abstract:

The majority of core engineering science courses in the School of Chemical, Biological, and Environmental Engineering at Oregon State University utilize a studio structure [1] where large lecture sections (100 - 350 students) are interspersed with smaller studio meetings (approximately 24 students). In studios, students work together in mostly 3-person teams, facilitated by trained graduate student teaching assistants (GTAs) and undergraduate learning assistants. Studios are designed to extend students' thinking and problem-solving techniques while simultaneously reinforcing core content and developing teamwork and communication skills. In its original manifestation, Studio 1.0, the activity relied on sequestered, worksheet-based problems. These problems helped students identify and practice key conceptual and procedural knowledge, and connect that understanding to lecture, but the activities were clearly limited in developing students' ability to connect the activity to professional practice and to develop value systems corresponding to the profession.

In the Studio 2.0 reform, our intent is to shift activity and re-situate learning by engaging students in *meaningful, consequential work* that directly and clearly relates to professional practice and desired professional attitudes and behaviors. The intent for these experiences is to provide a foundation for development of students' chemical engineering identity. Rather than attempting to direct students procedurally to a "correct" solution, a Studio 2.0 memorandum might explain a situation where a company is seeking to optimize a particular process and ask students to collaboratively decide on and perform calculations to make a design recommendation. Assessment is formative and immediate, focused on whether teams are "making progress" in grappling with the task. In this framing, as learners struggle with difficult concepts and may even sometimes fail to accomplish their short-term goals, they are continually positioned as engineers seeking meaningful progress towards a viable solution, rather than students following directions to get a grade.

In this paper, we present a four-hour studio progression where students work in teams in studio on a situated problem where they must determine the rate constant for a sugar reaction and, importantly, recommend a time for manufacturing to yield a minimum conversion. We have developed a technology-based solution where the teams are provided data that is generated by simulation. Each team has two reactors to work with and the simulation generates unique results for the team. Variation is added mandating the teams apply statistical reasoning in addition to utilizing their knowledge of kinetics. For some teams, the reactors are different, for others they are the same. The technology solution prompts students to respond to a set of prompts in which they report their findings and engineering recommendations and justify them. We report on how we use these collected data to better understand the way students are making choices in this ill-structured space, how students within the same team (with the same data sets) align, and how we can interpret the results to improve the activity design. Our ultimate goal is to automate these assessment processes to provide an instructor a broad understanding of the choices students are making that supports delivery and assessment of more authentic ill-structured tasks.

[1] Koretsky, M. D. (2015). Program level curriculum reform at scale: Using studios to flip the classroom. *Chemical Engineering Education*, *49*(1), 47-57.

531450 Reading and Repetition Using an Interactive Textbook for Material and Energy Balances

Abstract:

zyBooks creates interactive textbook replacements, and a zyBook for a material and energy balances course will be discussed. Features of the interactive textbook were highlighted previously, including animations, question sets, and auto-graded homework questions. Reading analytics have been presented previously with average reading rates of 87 and 93% for 2016 and 2017, respectively (n = 100 in 2016 and 88 in 2017). Now, new research questions are asked based on large sets of web analytics generated by students using/reading the interactive textbook. The high reading rates correspond to thousands of complete animation views, and new data will cover the most popular animations from the number of students re-watching an animation. Further analysis of studentsâ€™ attempts and success when completing auto-graded homework problems within randomized numbers and content will be presented. For the 2018 cohort, over 280 auto-graded questions were assigned to the students. Questions needing the most attempts before correct will be identified and analyzed. Finally, in addition to re-watching animations, students complete many attempts *after* answering the auto-graded questions correctly and earning their grade, i.e., performing self-guided repetition.

528915 Check Your Homework with Your Phone

Abstract:

Recently, computers have been used to deliver homework, quiz, and examination problems that were traditionally paper based. These electronic problems have several advantages for the instructor as well as the students. Advantages for students include instant feedback, multiple tries, and follow-up computer-based instruction to address any learner deficiencies. Less instructor grading and difficulties with students copying solutions manuals, along with the possibility of large problem banks are some of the advantages for instructors.

However, there are several disadvantages to computer-based delivery systems. Students are normally bound to the computer and an internet connection while they are working the problem. While this can normally be accomplished, it is certainly not as flexible as a problem on a piece of paper. Another disadvantage is that engineers are normally encouraged to clearly document their work and this step is less likely to be reinforced with a computer-based homework system.

The biggest disadvantage however, may be that students are less likely to efficiently engage the problems that are presented on a computer. This is especially true with traditional engineering education problems that tend to have several pieces of information, involve physical laws, mathematical manipulation and logical reasoning. When given these types of problems on the computer, some students will attempt to do more of the steps in their heads, usually becoming frustrated, instead of transcribing the material from the computer to a piece of paper to manipulate it. Obtaining a hard copy by printing it can sometimes be difficult if the problem is not formatted for printing.

To address these issues, a problem repository has been developed and continues to be populated with engineering problems that are made to be printed. Unlike traditional paper-based problems, each student receives a different version of the problem. In addition, the printed file contains a QR Code and website where students can check their responses with their phone or computer. Context specific instructional materials are provided on the website for students who need instructional support. Besides homework problems, this system lends itself to in-class practice and group problems especially for classrooms that are not equipped with computers.

Many of these problems have been tested in several chemical engineering classes including Energy Balances, Thermodynamics, and as a curriculum review in Chemical Process Design I.

The proposed system combines the flexibility and efficiency of a paper-based system with many of the advantages of a computer-based system and allows students to get immediate feedback utilizing their cell phones. The student interface for this problem system will be demonstrated in the session.

533334 Educational Augmented Reality Tools: Development, Implementation, and Assessment of Phase I

Abstract:

Pairing teaching and learning strategies with appropriate device and technology trends can greatly enhance the motivation to learn [1]. Through years of empirical studies, research on the use of computer or Internet in education has come to understand the relationships between learnersâ€™ characteristics and technology use, and has already shown some effectiveness of technologies in promoting learning outcomes [2]. Wearable technologies equipped with Augmented Reality (AR or Mixed Reality capabilities offer a lot of promise for education and training [3]. Such technologies have features that greatly align with the expectations of new student learners. At the same time, unlocking the full potential of these emerging technologies through proper tools and application implementation is essential for realizing the intended benefits [4]. However, compared to studies of more mature technologies in education, research of AR applications in education is in an early stage, and evidence of the effects of AR on teaching and learning appears to be shallow [5].

During the past years, in the context of EduART I (Transformative Educational Experiences 2016, a Texas A&M at Qatar grant), we investigated the use of AR to enhance the instructional experience of engineering students. A team of three faculty, one laboratory staff, two students, and an external programmer developed a visual support tool as a means to improve learning of chemical engineering students engaged in a specific laboratory activity while mitigating the challenges of laboratory work. The pilot demonstration of EduART I took place October-December 2017, where around 30 students, from multiple levels, participated. The experimental evaluation showed encouraging results, 43% showed significant and marginal improvements on the learning outcomes. A tremendous impact on the learning attitude was observed while impressive satisfaction and significant cognitive validity were captured in the self-surveys. To the best of our knowledge, there are very few similar efforts in engineering education and here we share the experiences gained during the development, implementation and assessment of EduART.

537060 Preparing Chemical Engineering Students for the Digitalization of Tomorrow – Integrating Modelling across the Curriculum

Abstract:

The advent of increased computational power, availability of more data and improved algorithms is allowing industry to "digitalize" operations. In essence, this means industry is able to use models to represent the plant accurately, use large amounts of data to determine the state of the plant, and use algorithms to optimize the plant performance on-the-fly. To prepare students for such a working environment, it is important to train students in the fundamentals of modelling. At present, modelling often does not go much beyond a flowsheeting exercise, which provides little understanding of the fundamentals underlying a model, and often has become a simple enter-data-and-push-button exercise. Modelling plays an important underlying role in almost all chemical engineering areas, from reaction engineering, physical transport phenomena to thermodynamics and of course process systems engineering, because the core laws, and chemical and physical phenomena, are expressed in equations. Introducing modelling as an integral part of a curriculum is therefore not only logic, but crucial in order to prepare students for the industry of tomorrow.

The Department of Chemical Engineering at University College London has successfully implemented an Integrated Engineering Program (IEP) where modelling is integrated across the entire curriculum, rather than being covered in a single course in the junior or senior year. To support the integration, PSE has developed and provided training material, exercises and tests through the Process systems engineering Academic Teaching Highway (PATH) initiative. The material is modular in nature and covers the typical fundamental topics from a modelling perspective and are supplementary to existing course material. We will reflect on the implementation and delivery of the new curriculum over the past four years, and how the PATH material has supported the students' learning, ultimately preparing them for a digital future.

537567 Wastewater Minimization and Energy Conservation Software Developed and Used in Teaching Process Systems Engineering at Vanderbilt

Abstract:

Russell F. Dunn, Ph.D., P.E.

Professor of the Practice

Department of Chemical and Biomolecular Engineering

Vanderbilt University

Nashville, TN 37235-1604

Ph: 615-343-4268

russell.dunn@vanderbilt.edu

Scott A. Guelcher, Ph.D.

Associate Professor

Department of Chemical and Biomolecular Engineering

Vanderbilt University

Nashville, TN 37212

Ph: 615-322-9097

scott.guelcher@vanderbilt.edu

Bryan R. Beyer

Adjunct Instructor

Department of Chemical and Biomolecular Engineering

Vanderbilt University

Nashville, TN 37235-1604

Ph: 615-343-4268

bryan.r.beyer@vanderbilt.edu

Wastewater Minimization and Energy Conservation Software Developed and Used in Teaching Process Systems Engineering at Vanderbilt

Significant efforts have been devoted to the development of process integration design methodologies over the past several decades. Two specific and widely-used developments in process systems analysis are the identification of optimal heat exchange networks and water allocation networks. These systems-based tools are widely applicable and are now essential elements of undergraduate process design instruction. This presentation will highlight the development of customized heat integration and water network software at Vanderbilt through senior capstone design projects. It will then be shown how these custom software packages are used in teaching process systems engineering in our design course and in applying process systems engineering in our senior unit operations laboratory course, when coupled with pilot plant data generated in the lab.

528546 Aspen Plus® Videos for Chemical Engineering Undergraduates

Abstract:

The University of Kansas (KU) Department of Chemical and Petroleum Engineering (CPE) has introduced ASPEN Plus® in the sophomore level Material and Energy Balance (MEB) course using videos and handouts created by an undergraduate who has taken the course. The ASPEN Plus® modules teach MEB students how to create a chemical process – the hydrogenation of benzene to cyclohexane in the program, including reactors, heat exchangers, distillation columns, and multiple recycle loops. The presentation will highlight how the videos were created using Camtasia Studio 9 screen capture software, the advantages of traditional learning via handouts and more innovative learning through videos, and benefits of a "student teaching students" learning experience. The goal of the speaker (junior CPE undergraduate) is to create ASPEN Plus® training modules for each course of the KU CPE curriculum. Thus, by graduation, a library of videos will be available for students to use in all the chemical engineering courses at KU.

533479 Problem Solving Skills When Using Student-Generated Problems That Reverse Engineer YouTube Videos

Abstract:

Homework problems generally come from textbooks, and many solutions manuals are available on the Internet. Students finding solutions manuals likely focus on copying the right answer and putting little effort on learning. Complex problem solving skill are in high demand for engineers. Here, students apply course concepts by reverse engineering YouTube videos and creating new homework problems and solutions. This research is examining rigor and problem-solving skills related to the YouTube problems. A quasi-experimental, treatment/control group design is in process. Rigor of YouTube problems is examined using the NASA Task Load Index (TLX), and problem solving skill are examined using a modified version of Problem definition, representing the problem, organizing information, calculations, evaluating the solution, solution communication, and self-assessment (PROCESS). Results from the first cohort from the 2017-2018 academic year will be examined and expand upon results presented at the 2018 ASEE annual meeting.

540773 2018 Outlook for Energy: A View to 2040

Abstract:

The Outlook for Energy is ExxonMobil's global view of energy demand and supply through 2040. It also highlights the dual challenge of providing affordable energy to support prosperity while reducing environmental impacts – including the risks of climate change. We use the Outlook to help inform our long-term business strategies and investment plans.
The Outlook anticipates significant changes through 2040 across the world to boost living standards, reshape the use of energy, broaden access to abundant energy supplies, and accelerate decarbonization of the world's energy system to address the risks of climate change.

- Global energy needs will rise about 25 percent over the period to 2040
- Electricity demand will nearly double in non-OECD countries
- Worldwide electricity from solar and wind will increase about 400 percent
- Natural gas will expand its role, led by growth for electricity generation
- Growth in oil demand will be driven by commercial transportation and chemical needs
- Efficiency gains and growing use of less-carbon intensive energy sources will contribute to a nearly 45 percent decline in the carbon intensity of global GDP

Energy is fundamental to modern life, and as the world's population approaches 9 billion people in 2040, we are challenged to improve living standards everywhere. Understanding the factors that drive the world's energy needs – and likely choices to meet those needs – is the mission of the Outlook. By sharing the Outlook with the public, we hope to broaden that understanding. Energy matters to everyone, and we all play a role in shaping its future.

541543 Energy Decarbonisation Scenarios

Abstract:

With the energy sector being associated with 2/3 of the annual greenhouse gas emissions, decarbonizing the energy is a key step to reduce their level. The National Determined Contributions (NDC's) announced so far by the participating countries do not meet the objectives of the Paris-climate agreement, as far as reaching a peak in emissions in the near future or decreasing to a level that is compatible with the stated long-term targets.

Using different scenarios for energy demand and supply, options for energy sector decarbonisation are evaluated. Meeting the 2-degree target requires a major transformation in the power sector, efficiency improvements – specially in the buildings and in the industry, and a revolution in transport towards fuel and modal shifts. With a significant increase in renewable energy, system integration and optimization will also be key in reducing the cost of the energy transition. Meeting more ambitious climate goals ("well below 2 degrees") will push further the technological innovation and implementation requirements throughout the entire energy value chain. A scenario aiming at zero net-emissions by 2060 is presented, along with a comparison to the 2-degree case.

542625 Fundamental Research Needs to Advance Energy Technologies

Abstract:

The U. S. Department of Energy (DOE), Office of Science, Office of Basic Energy Sciences (BES) supports fundamental research in chemical and materials sciences to provide the foundations for new energy technologies and to support DOE missions in energy, environment, and national security. This presentation will discuss opportunities for fundamental research to impact DOE's energy mission "to catalyze the timely, material, and efficient transformation of the nation's energy system and secure U.S. leadership in energy technologies" with a focus on the way we generate, store and use energy nationally. I will provide an overview of BES strategic planning over the past decade that identified priority research directions for advancing energy applications, highlight key scientific advances in these areas, and discuss some future opportunities for modern science, particularly chemical sciences, to accelerate the transformation of the U. S. energy portfolio.

540113 Using Team Building to Focus the Intersection of Diversity and Inclusion with Process Safety Practice

Abstract:

Within the past decade or so, process safety practice has undergone an important transition. While the importance of process safety has a long history of practice by many organizations, the profession of chemical engineering began to formalize such practices with the development of Risk Based Process Safety (RBPS) by the Center for Chemical Process Safety (CCPS) of the American Institute of Chemical Engineers (AIChE). RBPS has evolved over time, but its importance was recognized by OSHA early in the process, and OSHA included aspects of RBPS in its Process Safety Management (PSM) requirements. One of the important principles developed and practiced as a part of RBPS is that clear, relevant communication and training is critical to the effective practice of process safety. The importance of Diversity and Inclusion (D&I) is broadly recognized as important in our society, and in its most basic sense, diversity (bringing together people with different backgrounds) and inclusion (all those involved participating on an equal footing) are reflected in many aspects of RBPS. This paper discusses the intersection of D&I with process safety and particularly how training students to work effectively in teams supports these practices.

535890 Developing a Mindset of Safety in Students

Abstract:

Our process safety course â€œSafety, Health, and Loss Preventionâ€, which is an elective populated primarily by seniors, provides an in-depth look at specific elements of safety, some of which are also considered in the required courses of the curriculum. The overarching goals of the course are to provide students with a base level of knowledge of the safety regulatory structure in the US, industrial hygiene, and techniques for safer process design, and the ability to use appropriate models to quantify health and economic effects, toxicant dispersion, and damage due to fire and explosions. This is accomplished through a variety of techniques including in-class modeling exercises, expert guest speakers, and student led presentations as well as traditional lectures and homework assignments. The assessment of their understanding is based on several quizzes, exams, and their presentation.

One of our personal goals is for the students who complete a chemical engineering degree in our program to leave with a keen sense of the importance of safety, especially in the process industry. Our department works to develop that mindset of safety by incorporating small units relating to safety in several of our core courses, including a specific safety unit with activities in the Unit Operations Laboratory course, emphasis on safe behavior in all laboratory courses, and requiring EH&S in their senior design project. These units and activities provide a foundation on which they will build either in their careers or in our Process Safety, Health, and Loss Prevention (aka Safety) course. By including safety information throughout the curriculum, the students are reminded on a regular basis that safety must be a priority.

In this presentation, we will discuss our approach to this course as well as the departmental approach to incorporating safety throughout the curriculum.

527167 Process Safety Instructional Enhancements Implemented in a Two-Semester Chemical Process Design Course

Abstract:

An ABET review of our chemical engineering program in the first year of the authorâ€™s teaching appointment at Texas A&M University-Kingsville (TAMUK) spurred a comment regarding the need for greater process safety instruction in our curriculum. Our department faculty indicated this instruction would best fit into both the unit operations laboratory two-course sequence and the chemical process design two-course sequence. The author, who is the instructor for the chemical process design course sequence, has, over the last two years, incrementally developed and presented new units on chemical process safety for both of the design courses. The emphasis of the new units has been on four different areas. The first of these focus areas is identifying and understanding reactivity, flammability, and toxicity hazards of the chemicals studied in various course design problems, including the group-focused senior design project in the second course. The second focus area is risk-based process safety management systems commonly implemented by refinery and chemical process corporations. The third focus area is on review and analysis of actual chemical accidents, such as through event reconstruction videos provided by the Chemical Safety Board. The fourth focus area is on understanding and creating HAZOPs tables for limited portions of the process each group is developing for their senior design project. Students are evaluated on their understanding of this material using quizzes in the second design course, as well as on the extent to which these principles are included in their senior project chemical process design. The author uses material obtained at the CCPS Faculty Workshop on Chemical Process Safety from June 2016, as well as material from Crowley and Louvarâ€™s *Chemical Process Safety* text and Turtonâ€™s *Analysis, Synthesis and Design of Chemical Processes* text, to develop these instructional units. This paper focuses on presenting the methods used for instruction and the successes and challenges indicated by student work in grasping this course material over the prior two years of chemical process design courses at TAMUK.

537514 Integration of Process Safety Experience in Research Project into Undergraduate Process Design Courses

Abstract:

The undergraduate, senior process design course at The Ohio State University comprises a series of two 4-credit courses, namely: Chemical Process Principles: Economics and Strategy and Chemical Engineering Design and Development. The first course (Process Principles) introduces and teaches senior chemical engineering undergraduate students a core set of process design principles, including process safety evaluation, economic evaluation, optimization theories and heat integration. The second course (Design and development) exposes students to managing, working and communicating on a real world (industry/academia sponsored) open-ended chemical engineering project wherein they apply the skills learned from the first course towards completing the project. This two-tiered structure gives students an in-depth learning opportunity to learn skills in one course and reinforce them by application to a real-world project in the second.

This talk will focus on the process safety facet of the two courses. Specifically, over the last two years, the presenting author and the instruction team have restructured the process safety lessons taught in the first course (skill building) to revolve around a specific chemical looping pilot plant constructed by the instructor and his team. Concepts of process safety analysis, safety relieve valve sizing, inherently safer design are directly exemplified in the pilot plant design making a clear translating class room lessons to real-world applications. Though this approach engages students by rooting the class on a specific and detail process, such approach requires additional challenges it requires the students to familiarize themselves with the process concept, learn how to interpret control and mechanical specification documents, and decipher piping & instrumentation diagrams. However, with this approach, students can be better prepared to enter industry. The learnings from the first course are reinforced in the second, where the students apply PHA and risk assessment techniques to their open-ended projects. The present paper will initially discuss the experience in developing material that incorporated the chemical looping pilot plant development in the process design teaching material. Experience gained from the last two years of instructor this course in regards to student response and course refinement are discussed. Finally, examples and perspectives of the applications PHA and risk assessments from the first course used in practic by students in their second course will be presented. An overall perspective on specific changes that will be implemented in Fall 2018 and direction of further safety material will be shared.

534917 Integrating Chemical Process Safety across the Curriculum

Abstract:

Process safety education and training at US colleges and universities is currently insufficient to meet industry needs.[1] Failure to prepare students with essential process safety training can and does lead to catastrophic and financially devastating events. In addition, failing to include mandated process safety in the curriculum can cause a program to lose ABET accreditation. Loss of accreditation means loss of competitiveness and will result in fewer enrollees, which could add up to millions of dollars lost in annual tuition per year. The goal of this project is to integrate process safety across the Northeastern University (NEU) Chemical Engineering (ChmE) curriculum to address these needs and minimize these risks.

Industry has identified eight necessary chemical process safety learning outcomes as specified by the Safety and Chemical Engineering Education (SAChE) Committee[2]: These outcomes span from an understanding of past events, to the hazards and impacts of chemicals and releases, and move towards understanding how to assess the hazards and risks associated with a process and to design or mitigate against them.

Through this project, the NEU ChmE curriculum was updated to include these eight outcomes, integrated across eight core ChmE courses, taught by 14 instructors, in order to evaluate the impact of the curriculum change. The student perceptions of their respective skills were measured before and after the redesign and their scores on assignments/quizzes/tests that related to process safety needs were also measured. In addition, this project produced the framework needed for future collaborative initiatives in chemical engineering education. Faculty annual reviews now include specific and measurable educational goals. A review system for faculty to measure their education outcomes and reflect on, and propose future changes has been implemented, and an annual faculty education retreat has been established to develop department wide educational goals. In addition, students have created an electronic portfolio to showcase the evidence of their effective progress.

1. Luo, Y., Westmoreland, P. R., Alkaya, D.,Alves da Cruz, R. V., Grossmann, I. E., Provine, W. D., Silverstein, D. L., Steininger II, R. J., Talbot, J. B., Varma, A., McCreight, T., Chin, K, Schuster, D. *Academia-Industry Alignment: Expectations about New Graduates, an NSF-Sponsored Study led by the American Institute of Chemical Engineers*; 2015.
2. Safety And Chemical Engineering Education Committee, (SAChE), [cited 2017; Available from: http://www.sache.org/SACHEGuidelinesForABET.pdf].

532442 Where Does Process Safety Fit into the Chemical Engineering Curriculum?

Abstract:

This talk will discuss some of the options that all chemical engineering programs should consider. Should process safety be integrated through the curriculum or mostly be focused into a single separate course? Should the greatest emphasis be placed on regulations, computations, or case studies? How does process safety link to other hard to place topics such as technical communications and professional ethics?

538833 Integrated Humor in the Engineering Classroom Using Isomorphic Mapping to Improve Learning Outcomes Via Increased Germane Cognitive Load

Abstract:

Numerous claims that humor can improve education have been made in various areas. However, there is precious little direct evidence of this, particularly in technical areas such as engineering. A method to formulate such humorous/relatable examples is described here that is simple for technical educators to use in educational interventions. Using a cognitive load survey, adapted from previous survey instruments, we characterized the cognitive load components of students in response to traditional and humorous/relatable educational interventions in two chemical engineering classes. Early results suggested that addition of humor was not effective. We hypothesized that humor distracted engineering students by increasing the extraneous cognitive load (the mental effort used in processing material not relevant to the educational lesson. To this end we integrated humor into the educational lesson to reduce the extraneous cognitive load. We hypothesized that a humorous/relatable example integrated into the educational lesson would increase the germane cognitive load (the mental effort used to observe patterns in the new material and facilitate learning). We hypothesized that a humorous relatable exmples would increase the germane cognitive load because the students could relate to such an example. If this example was sufficiently similar in structure to the actual technical lesson (isomorphic) then knowledge transfer to the actual lesson would occur. This adapted cognitive load survey produced self-consistent results that were used to determine what type of humorous/relatable intervention could improve the germane cognitive load associated with improved educational outcomes. If the intrinsic cognitive load is relatively high, a humorous/relatable example integrated into the educational lesson can improve germane cognitive load. However, the example must be isomorphic in structure to the associated educational lesson to achieve such results. If the intrinsic cognitive load is sufficiently low, then increased germane cognitive load can be achieved without such an intervention. Humor itself is not actually required as long as the humorous/relatable example is isomorphic and integrated into the educational lesson. However, humor often results from the incongruity of mixing a technical and non-technical concept that are isomorphic. Humor can add the additional benefits of increased student engagement and instructor immediacy.

534064 Making Interactive Simulations and Screencasts More Interactive

Abstract:

To more fully engage students in learning, we have prepared interactive simulations that lead students through step-by-step procedures and screencasts that pause and require students to answer questions based on what they have just watched. Most of these simulations were programmed in Mathematica, and they are available on www.LearnchemE.com and on the Wolfram Demonstration Project site. A few web-based interactive simulations that give students a grade were also prepared and are available on the LearnChemE site. The software PlayPosit® was used to prepare the interactive screencasts by inserting questions into a previously-prepared screencast.

We currently have more than 160 interactive simulations (www.LearnCheme.com) that allow students to change parameters using sliders and buttons and observe how a system responds. The new simulations to be discussed are based on simulations reported by Bansagi and Rogers [1] for ternary phase diagrams and liquid-liquid extraction. They reported that student grades improved and the gap between poor and good students decreased when their simulations were used. In these simulations, each step requires students to locate a line or a point on a graph or input an answer, and the simulation than provides the correct answer. A hint is available for each step, but once the correct answer is shown, the student cannot go back to redo that step. They can, however, start over by selecting a new problem, which starts with different parameters and/or starting conditions. To date, ten step-by-step interactive simulations have been prepared for use in thermodynamics and separations courses, and a number of additional ones are being prepared. Although they are programmed in Mathematica, they can be used with free software available on the Wolfram Demonstration Project website (http://demonstrations.wolfram.com/about.php)

We have previously prepared more than 95 Interactive screencasts that consist of one multiple-choice conceptual question. Students click a link in the screencast to select an answer, and they are then directed to another screencast where they receive feedback for their response. However, these new interactive screencasts deliver content and then the screencast pauses and a quiz pops up that contains multiple choice questions (and other formats) about the material just viewed. Any number of questions can be inserted this way, and formative feedback is provided to learners. These screencasts with questions require the PlayPosit® software, which also allows the instructor to monitor student responses to the questions. The answers are automatically graded and the grades recorded directly into a learning management system.

Reference

[1] Bánsági Jr and T.L. Rodgers, Graphic web–apps for teaching ternary diagrams and liquid–liquid extraction, *Education for Chemical Engineers*, 22, 27-34 (2018)

536532 Introducing Contemporary Topics in a Transport Phenomena Course through Michigan-Themed Projects

Abstract:

Chemical engineers contribute to solutions in an increasingly diverse set of areas. As such, it is essential for students to be trained to work effectively in cross-disciplinary teams and to hone their leadership, communication, and interpersonal skills. One method to accomplish this is project-based learning and, in particular, projects based on contemporary topics with a local scope. Such projects can be truly motivating to students since they can perceive their impact on the student's communities, families, and themselves. Furthermore, this type of project can emphasize the effects that engineering solutions may have on a variety of contexts – economical, societal, and global, to name a few.

Transport phenomena represents one of the cornerstone topics in the chemical engineering curriculum. At many universities, transport phenomena is taught among several courses; for example: fluid dynamics, heat transfer, and mass transport/separations. Some programs have a modern "Transport Phenomena" course that incorporates modeling and computer simulations to expand on the basics of transport already discussed in previous courses. This more focused class can be challenging for undergraduate students because of its mathematical rigor and the difficulty to link the theoretical framework to applications in a direct way.

We created a set of projects for the senior-level "Applied Transport Phenomena" course at Kettering University, which were assigned as a final term project for the course. These projects were all related to topics associated with Michigan and drew on a variety of issues – including environment, safety, health, ethics, society, and politics. Project titles were as follows: "Dioxane Plume in Ann Arbor Groundwater," "Michigan Lakes Freezing Depths – Safety and Environmental Effects," and "Corrosion and Leaching of Lead in Pipes – Flint Water Crisis." For each project, students were first tasked with completing a thorough literature review to understand the issue and to investigate the related history. They were asked to model the key transport phenomena related to each project – mass, heat, and momentum transfer were involved in each case. The projects allowed students to draw on theory developed in class and to apply it to a specific problem. Project constraints were altered to study the effects on the rate of transfer unique to each project; for example, the rate of leaching from pipes was estimated based on concentration of the water or based on the thickness of lead coating the pipes. Students worked in teams and presented their findings orally and in a written report.

Students completed surveys meant to understand their perspectives as well as to probe whether the projects enhanced their understanding of the course material and how it is applied. Details of the projects and a summary of student feedback will be presented.

533793 Teaching Energy and Sustainability Using Only Active Learning Pedagogies

Abstract:

Knowledge of Energy and Sustainability are important for undergraduates to learn about before moving into the work force. The challenge with such fields is the information is too vast and intricate to be justifiably conveyed in a university course through standard lectures. Interaction and engagement with the information that yields the highest level of understanding is critical. In this talk, I discuss my course offering entitled â€œMaterials for Energy and Sustainabilityâ€, in which students engage with the course content using only active learning pedagogies. The learning activities include: Student-led discussion, small group problem-solving, peer-judged debates, peer-to-peer lightning talks, as well as competitive games. I will draw from the outcomes of these learning activities to show how students gain excellent understanding of the critical aspects of Energy and Sustainability, and how not only more knowledge can be acquired as compared to lectures alone, but also better mastery of that knowledge, in a high-energy and engaging learning environment. These activities and their positive outcomes generalize to other upper year elective courses that aim to introduce students to broad technical fields.

A Blended Textbook-Free Chemical and Biochemical Process Safety Class

Abstract I.D.: 532388

Daniel Forciniti

Kielhorn Research Laboratory, Chemical and Biochemical Engineering Department, Missouri University of Science and Technology, Rolla, Missouri 65409, United States

ABSTRACT

The Chemical Engineering Department of Missouri S&T has taught a chemical process safety class for more than twenty years. However, students pursuing a biochemical engineering emphasis program were not required to take that class. The department started a new curriculum two years ago that include a safety class for everyone. A new class was developed that includes elements of both chemical and biochemical process safety. There are several unique features of the new class that I would like to share: 1) it is offered in blended mode with two in class meetings per week followed by a series of online discussions; 2) it uses one of the design projects in senior design as a case study. Therefore, the material covered in class as well as the topics of the discussions are concentrated on a particular process; 3) a textbook is not required or recommended. Instead, the students use primary sources as well as information freely available on line published by various companies; and 4) this is a communications intensive class. The discussions are used to both improve the written communication skills of the students as well as to dig deep into a selected number of topics covering both chemical and biochemical processes. The discussions start with a brief posting about a manuscript that the students are asked to read. That is followed by a discussion that uses several prompts developed by the instructor. Student participation in the discussions have been excellent, providing valuable information to the class as well as to the instructor.

INTRODUCTION

Process safety may be defined in a variety of forms. For example, CCPS defines it as "a discipline that focuses on the prevention of fires, explosions, and accidental chemical releases at chemical process facilities. It consists of the protection of people and property from **episodic and catastrophic** incidents that may result from unplanned or unexpected **deviations in process conditions"**. (Guidelines for Auditing Process Safety Management Systems, 2nd ed. (New York, CCPS, 2011xxi). In Canada is defined as "a blend of engineering and management skills focused on preventing **catastrophic accidents**, particularly explosions, fires, and toxic releases, associated with the **use of chemicals and petroleum products"**. (M. Macza, "A Canadian perspective of the history of process safety management legislation (Cologne, Germany, 2008, 12/2). The Energy Institute's expansion of the CCPS definition is: "Process safety is a blend of engineering and management skills focused on **preventing catastrophic accidents** and near misses, particularly structural collapse, explosions, fires and toxic releases associated with **loss of containment of energy or dangerous substances** such as chemicals and petroleum products. These engineering and management skills exceed those required for managing workplace safety" (www. Energyinst.org/technical/safety/process-safety). Therefore, process safety is focused on preventing deviations in process conditions that lead to loss of containment of energy or dangerous substances that caused catastrophic accidents.

The fundamental concepts of process safety were developed by industry and not by regulating authorities, as is commonly assumed. Although process safety is mostly focused on the chemical and petrochemical sectors, it has extended into bioprocessing plants where biological hazards are added to chemical and physical ones. A broad definition of bioprocessing includes facilities dedicated to the production of pharmaceuticals, biochemicals, chemicals, fuels and food and food additives. The scale of bioprocesses plants is equally broad. It includes very large scale processes like the production of ethanol (having 500,000L reactors) or citric acid (with a demand of 1M tons/year) to mid-size plants like the ones for the production of antibiotics or enzymes (100,000L reactors) and small size plants like the ones producing mAb (10,000 to 25,000 L) or vaccines (2,000 to 5,000L). Examples of biological hazards include the release of the microorganism (biohazards), pharmaceutical active products, waste solid or liquid, emissions to air and wastewater discharges.

Process safety concepts are often covered in several classes across the Chemical Engineering Curriculum instead of in a course dedicated to the topic. The chemical engineering department of Missouri S&T started requiring a process safety class to those students affected by catalog year 1994/95. Pioneering that effort was Dr. Neil Book who has retired a few years ago. Three years ago, the department started offering a Biochemical Process Safety class for the Biochemical Engineering Emphasis students. Recently, the two classes have been combined into a single Chemical and Biochemical Process Safety class.

The Student Learning Outcomes of new class are: SLO 1) to identify and quantify the different hazards associated with chemical and biochemical processes, SLO 2) to assess, quantify and prevent risk in Chemical and Biochemical Processes and SLO 3) to develop a plan to mitigate an event.

The students learning objectives and their pairing with SLOs are:

1. (SLO 2)Explain biochemical process to different audiences (to prevent fear of the unknown. i.e. demystify scientific research and production of biochemical processes to different audiences to counter SciFi and horror media). Assessment: discussion post.
2. (SLO 3) Compare and contrast Global regulations (focusing on US and EU). Assessment: discussion post.
3. (SLO 2) Identify/explain potential hazards in the food industry. Assessment: discussion post.
4. (SLO 2) Identify/explain potential hazards in refineries with an emphasis on biodiesel and biofuels. Assessment: discussion post.
5. (SLO 2) Identify/explain potential hazards bioprocessing industry. Assessment: discussion post.
6. (SLO 1) Compare and contrast dust and liquid originated fires and explosions and their consequences. Assessment: i) Homework, ii) 1/2 of a test.
7. (SLO 1) Assess the effectiveness of different relief devices. Assessment: i) homework, ii) 1/2 of a test, iii) discussion post.
8. (SLO 1) Apply mathematical models to predict the path of aerosols. Assessment: i) homework
9. (SLO 1) Compare different risk assessment approaches. Assessment : i) homework-discussion post (--event trees, fault trees, HAZOP, LOP) ii) ½ of a test
10. (SLO 1) Assess risk in chemical and biochemical processing equipment. Event trees, fault trees, HAZOP, LOP. Assessment: i) group discussion post, ii) Final exam.

The highlights of this new class are: 1) it focuses on a design theme, 2) an entry test is administered the first week of classes, 3) it is a blended class consisting of two hours lecture in class and one online weakly discussions, 4) it covers chemical and biochemical process safety and 5) it is a textbook free class. The process design and the process safety instructors need to coordinate their efforts to use the time of both classes more efficiently by not overlapping on process safety topics. Process safety is a senior level class.

Therefore, the students taking the class should have a good foundation of thermodynamics, transport phenomena, separations, chemical kinetics and reactor design. The entry test is meant to assess the knowledge of the students on these topics and to offer remedial material for those who need a refreshing. This class may move completely on-line in the near future. The class is currently offered in blended mode with two meetings in person per week plus one on line meeting. The online meetings are focused on various discussion topics chosen by the instructor. The class needs to satisfy the needs of both regular and biochemical engineering emphasis students. Although there is an overlap between chemical and biochemical processes safety in several topics such as like lack of containment (fires and explosions), relief devices, risk assessment, and aerosols, the biotechnology industry has unique concerns about process safety related to pathogenic organisms or the release of toxins. I have selected a set of topics that can be understood by a person with little biology knowledge to address those special needs. Finally, I decided not to use a textbook for this class. Although I value the process safety books available to the students, I decided to use this class to introduce the students to learn material as in the real world, where information may be available from government originated documents, vendors or the scientific literature.

The class consists of the following units:

Unit 0 (2 weeks). Prerequisites Review

- a. Introduction of Case Study
- b. Thermodynamics
- c. Fluid and Heat Flow
- d. Process Units
- e. Probabilities and Statistics
- f. Flowsheets

Unit 1 (2 weeks). Introduction to Process Safety

- a. Chemical and Physical Process Safety
- b. Biological Process Safety
- c. Bioprocessing incidents and releases
- d. Current Biosafety guidelines by Agency

Unit 2 (1 ½ week). Identification of Hazards

- a. Physical Hazards
- b. Chemical Hazards.
- c. Dow's Chemical Exposure Index
- d. Biological Hazards. Survival of Microorganisms on Inanimate surfaces, A. Kramer and O. Assadian in: Use of biocidal surfaces for reduction of healthcare acquired infections, 2014, pp 7-26.

Unit 3 (1 ½ week). Chemical and Biological Aerosols

- a. Aerosol Physics
- b. Agent Considerations
- e. Process Considerations
- f. Environmental Considerations
- g. Ghosh B., Lal, H., Srivastava, A. Review of bioaerosols in indoor environment with special reference to sampling, analysis, and control mechanisms. Environment International, 85: 254-272 (2015)

Unit 4 (2 weeks). Lack of Containment, Fires and Explosions

- a. Definitions
- b. LOC and Flammability Diagrams
- c. Inerting or Purging
- d. Ignition
- e. Static electricity. Ignition hazards caused by electrostatic charges in industrial processes. M. Glor and Thurnheer, P. 2015.

f. Releases
g. Source Models
h. Fires
i. Explosions
j. Dust Explosions. https://www.osha.gov/pls/oshaweb/owadisp.show_document?p_table=STANDARDS&p_id=9874

Unit 5 (1 ½ weeks). Introduction to Relief Systems

a. Location of relief systems
b. Relief types and characteristics
c. Relief systems
d. Design of knock down drums (http://www.api.org/~/media/files/publications/whats%20new/521%20e6%20pa.pdf

Unit 6 (1 week). Safety in Bioprocess Facilities Design

a. Heating and ventilation systems
b. Containment
c. Waste and Waste Treatment. Nadal et al. Health risks of the occupational exposure to microbiological and chemical pollutants in a municipal waste organic fraction treatment plant. Int. J. Hyg. Environ. Health, 212:661-669 (2009)
d. Building and site security
a. Cleaning, inactivation and sterilization
b. Air and gas emissions

Unit 7 (2 ½ weeks). Risk Assessment

a. Failure Rate and Reliability
b. Boolean Algebra
c. LOPA
d. HAZOP and Safety Reviews
e. Fault and Event Trees
f. Identification of major accident hazards in industrial biological processes. Moreno, V.C., Giacomini, E., Cozzani, V. Chemical Engineering Transactions, 48(2016) 679-684.

The references at the end of each unit are the material the students are asked to read prior to their participation in the discussion.

Design Theme. The design theme for the last semester I have taught this class (Fall 2017) was a biodiesel plant using canola oil as the raw material. The topic was briefly reviewed in the process safety class. It was expanded to include heterogeneous catalysis. Potential hazards were discussed. They included the use of toxic materials like methanol and sodium hydroxide as well as the possibility of overpressure in the distillation columns.

Entry Test. The entry test includes topics that are relevant to the topics covered in the class. They include mass and energy balances (including differential balances), separations, thermodynamics as well as bio specific topics like pathogens. The test was built into canvas and it is, for the most part, self-graded. There are some assay type of questions that need to be graded by the instructor. A couple of sample questions are shown in Figure 1.

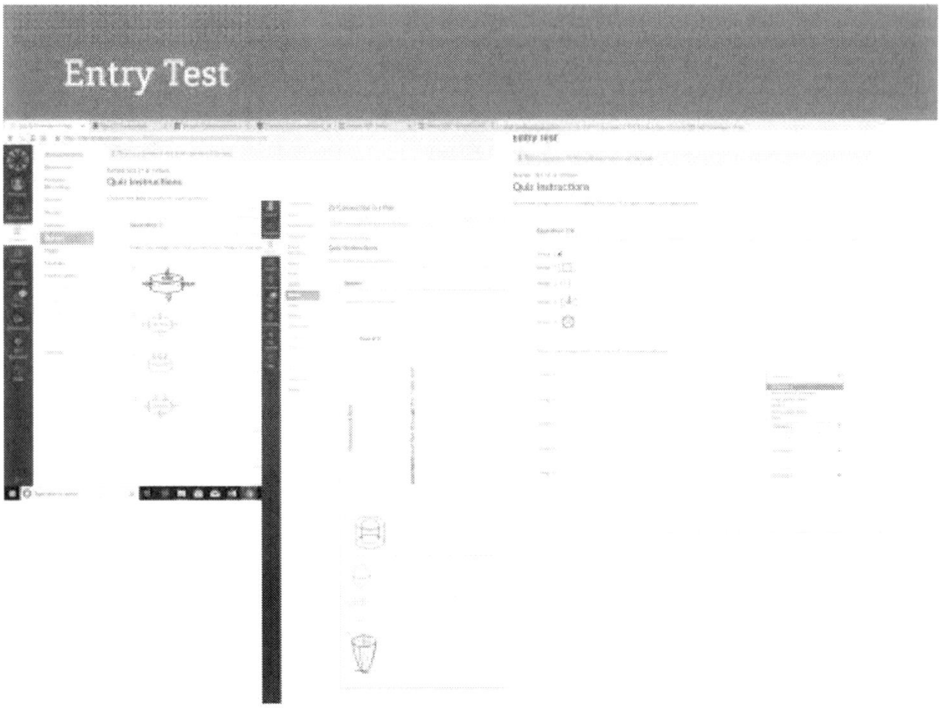

Figure 1. Sample entry test questions. The ones on the left panels continue with subsequent questions leading to the differential equation (and boundary conditions) that describe the phenomenon. The one on the right panel quizzes the students about their knowledge on flowsheets symbols.

The results of the test are used to design the remaining classes for the first two weeks. Some of the material is reviewed in class but some is covered with short videos (5 to 10 minutes each).

Blended Class. The class meets on Mondays and Wednesdays but on Fridays, discussions are conducted on line. Participation in the discussions has been excellent with an average of 3 participations per student per discussion. The class is writing intensive; therefore, both content and language assessments are part of the rubric. The following are a few descriptions of organization of individual as well as group discussions.

<u>Comparison between EU and US process safety regulations.</u> Read the ATEX directives. Use the regulations to evaluate how strict the regulations are in the U.S. compared to the various European countries, including consistencies and inconsistencies in their application, particularly in biofuel vs. oil and gas.

<u>Contamination of Surfaces by Microorganisms.</u> Start the discussion by choosing a microorganism type (bacteria, fungi, viruses, etc.) that is also used in biotechnology. Identify the microorganism as Gram positive or Gram negative. Discuss the easiness/difficulties for disinfecting surfaces contaminated by that particular microorganism. Continue the discussion analyzing at least one of the factors (as they relate to process) that would affect the survival of the bug (humidity, temperature, surface roughness, biofilms).

<u>OSHA Grain Silo Regulations.</u> Focus on regulation 1910.272 App A. Explain what a hot work permit is and give at least one example of it during the maintenance of a biodiesel facility. A link to OSHA can be found here: **<u>Link</u> (Links to an external site)**. Then focus on Dust Emission Control and Filter Collection. Compare and contrast two different approaches to maintain the amount of dust accumulated below the tolerance limit, explain pros and cons.

Static discharges. Each of you is assigned a different manifestation of static discharges: sparks, brushes, corona, and cone discharges. Start the discussion with the main features of your assigned discharge as well as identifying the main causes (in processes) of that discharge. This is a good source **Link (Links to an external site).** Continue the discussion by posting on others' opinions/questions about the efficacy for that particular manifestation to ignite different flammables as well as preventive measurements.

Aerosol sampling. Each of you will play the role of a vendor for a specific aerosol collection system or enumeration technique. You must start the discussion with a brief description of the technique assigned to you, enumerate advantages and disadvantages and its cost. After the initial post, pick up a competitor and highlight the advantages of your device compared to the competitor.

Design and placement of reliefs followed by the design of knock down drums. Discuss the placement (and type) and installation of at least two relief devices in a biodiesel plant based upon the recommendations of APR-521 (First due date). Find and post the proper design equations for the devices based upon a mid size production plant, discuss the assumptions leading to those equations (Second due date). Solve the equations to obtain the necessary area in squared inches. The release cannot be vented directly to the atmosphere. Therefore, you have to size a knock down drum. You may use the "exact" calculations or use the approximate solution discussed in class this week. The discussion may include (but not limited to) critiques about the sizing approached followed, the location of the devices and the selection of the type of device.

Risk Assessment and HAZOP. This is a group discussion. The class was divided into four groups of seven students each. The project leader was chosen by the instructor. The decision was made based upon the extent to which each student participated in the individual discussions. The other students were randomly chosen. The first task assigned to the project leader was to assign the following roles to the remaining students in her group: a) process engineer, b) safety engineer, c) operator technician, d) maintenance engineer, e) automation engineer, and f) project engineer. Initial Post: The team leader selects two (2) nodes for a reactive distillation biodiesel plant and describes the roles of the different members of the team. The team then brainstorm process parameters, deviations, causes, consequences (including evaluation or risk, i.e. frequency and severity), and action required. At the end of the discussion, the team submits a HAZOP form (from my notes).

Biohazards. Biohazards may have severe consequences to the operators and to the surrounding areas. Although deviations and the consequent loss of containment of some bioprocesses are relatively harmless, some are extremely hazardous. For example, loss of containment in defense laboratories and in the production of vaccines against highly contagious and virulent strains. In general, the following information is relevant about a particular biological agent: 1) pathogenicity and mode of transmission, 2) infectious dose, 3) virulence, 4) host factors, 5) sensitization reactions, 6) availability of vaccines and 7) environmental impact. These topics are discussed in class. The students are introduced to the various biosafety levels and acronyms and their relationship with specific pathogens.

Textbook Free. Textbooks are common sights in undergraduate engineering classes of American universities. There are classes in which the textbook is followed closely, some in which it is not followed at all, and anything in between. However, a textbook is always required. Some studies show that students feel more comfortable in "textbook based" classes than in classes without one, in spite of the high cost of textbooks. In addition to the high cost, in today's technological word it is impossible to cover current knowledge in a single monograph and thus it is becoming more important to explore alternatives to a textbook in engineering classes. The reading material for this class are given to the students through Canvas with the assistance of the library to sort out copyright issues. The material consists of (1) A number of scientific papers and vendors' material for each of the main core topics, (2) A series of power point slides/notes that are used to support the lectures (the slides are not used as the main teaching method in this class), (3) Detailed reading guidelines. By doing this, the instructor provides the structure that the students seem to need to be able to start thinking outside the box and learn directly from primary sources.

Summary

Students' reception to this class was mixed. These are a few excerpts from the students' comments. "I do like how this class is more discussion based than the other risk assessment course that a lot of students are taking. I believe that safety assessment is a conversation based on peoples (sic) opinions and should expand into a discussion". "The discussions didn't seem like the best way for students to learn material. As i (sic) said before, I feel like students learn the best by working examples and discussions were not the way to do this. Also as I said before, it was concerning how low averages were on the tests, especially for a process safety class which students have seemed to do very well in prior to this semester". "I do like how this class is more discussion based than the other risk assessment course that a lot of students are taking. I believe that safety assessment is a conversation based on peoples (sic) opinions and should expand into a discussion". I will be glad to share with anyone the list of references that I use in this class as well as discussion rubrics and complete examples of online discussions.

530335 Revealing the Decision-Making Processes of ChE Students in Process Safety Contexts

Abstract:

Process safety has always been important within the chemical engineering profession. Chemical engineering companies use a variety of methods to ensure that their employees have an understanding about and are able to follow process safety protocols. These methods include e-learning focused on instructional systems design, process safety management techniques such as internal audits and checklists, and hazard identification training. Organizations such as AIChE and SAChE also offer safety courses online for individuals working within this industry. However, most of these programs and methods focus on recognition of process safety hazards and do not emphasize the importance of the reasoning that surrounds process safety decision making. This same problem also occurs in undergraduate classrooms where students are taught to recognize safety hazards and warning signs, but instruction is not necessarily focused upon the thought process that accompanies a process safety decision.

This absence of work to understand the reasoning that individuals use to make process safety decisions is of particular importance since work done by Sadler and Zeidler (2005) has shown empirically that students approach socioscientific decisions using some combination of three distinct forms of informal reasoning: rationalistic, emotive, and intuitive. Rationalistic informal reasoning involves the use of logic and weighing options when considering a decision. Emotive informal reasoning includes evaluating the impact the decision might have on others involved and exhibiting empathy by trying to understand the feelings of others. The last form of informal reasoning, intuitive, relates to following oneâ €™s â€œgut feelingâ€ or first reaction to a situation.

In this study, five senior chemical engineering students described out loud their reasoning when making decisions about seven distinct process safety scenarios. The process safety scenarios represent dilemmas that individuals might face in industry and ranged from making choices about whether to report chemical spills to governmental agencies to determining if staff should stay in place at a chemical plant when there is an impending severe weather event. Each student was asked to elaborate on what role they believed that rationalistic, emotive, and intuitive informal reasoning served in their decision making process. These think-aloud sessions were transcribed and analyzed qualitatively to understand the ways in which engineering students reason through hypothetical decisions related to process safety. This presentation will discuss the results obtained from this study and observations made on which informal reasoning or combination of these approaches students use when making process safety decisions.

539670 A Curriculum-Wide Strategy for Student Teaming Skills Development in Engineering

Abstract:

Teamwork is an essential component of successful engineering work and often includes teaming with members of different disciplines and professions. Several senior-level instruction and activities on team skills were piloted in a School that houses Chemical Engineering, Bioengineering and Environmental Engineering programs. The end goal is to evaluate and improve activities for a curriculum-wide teaming skill development strategy for ongoing work to revolutionize the undergraduate learning environment. Optimally, our strategy provided students with a structure conducive to practicing inclusive teaming along with guidance and training to improve their skills. The teaming instruction and activities were delivered in lab, studio and design courses to at least 185 students of all three disciplines. Students were provided the opportunity to reflect on the effectiveness of the instructions and activities, and make suggestions based on their team experience. A focus on conflict resolution strategies was undertaken, and the students reported that learning about conflict types and resolution emphasized to them the need for teamwork skills. At the completion of the year, the framework for a curricular wide teaming skills strategy was presented to students, and their feedback solicited. The quantitative and qualitative results from student feedback on piloted activities will be reported. A framework for curriculum-wide development of teaming skills will be presented, as well as assessment approaches for individual team skills in large course formats. Lessons learned will be shared to show successes and areas of improvement for student teamwork skill development.

537689 Applying Engineering Optimization Principles to Engineering Education: Optimization of a Student Project Experienceâ€™s Design and Implementation

Abstract:

Engineering is the field of optimization and as such engineering faculty are uniquely equipped to apply an optimization perspective to education. Here we consider both the constraints with current best practices toward the objective of obtaining long-term student success as applied to an engineering course group project experience. Specifically, we apply the global education optimization objective of positively impacting students to develop skillsets that enable students to be successful in both career and in life. Student success is defined globally as both overall wellbeing (emotional, mental, social, and physical) in addition to engagement in the workplace, which is consistent with recent long-term studies. The constraints of the education environment are generally defined as 1) time, 2) financial resources, and 3) student/faculty motivation and abilities. The impact of this approach was assessed following IRB approval and the results of a multi-year study are reported herein. Overall, we find great value in using optimization principles to help individual instructors as they continually work to improve the educational framework for their students.

528980 An Interdisciplinary Research Program for Undergraduate within the Functional Materials and Manufacturing Institute at University of South Florida

Abstract:

Upper level undergraduate students often need reinforcement of basic skills, previews toward more challenging topics, a mindset toward design, guidance toward career options, and a variety of opportunities to develop and practice communication skills. At the Functional Materials and Manufacturing Institute (FMMI) at University of South Florida (USF), these topics are simultaneously addressed through research in functional materials. Research is identified as a method of engagement and retention of students. Functional materials are prevalent throughout the modern world and are constant reminders of advances from engineering. Thus, research experiences in functional materials, with a mindset toward application/manufacturing, is an outstanding topic to engage emerging researchers with a wide variety of interests. Most emerging researchers in the FMMI research program are upper level undergraduates in chemical engineering, but the research program has included high school students, lower level undergraduates, and K12 teachers. In this contribution, the overall program for emerging researchers at FMMI will be discussed. Specifically, the non-research elements will be highlighted with an emphasis on junior/senior undergraduate researchers. These elements include career guidance, technical writing and presenting, and aspects of safety and ethics in an interdisciplinary environment, which permits creativity and variety in the research experience. With a current NSF REU site grant, correlations are made between applicants majoring in the chemical engineering selecting projects advised in a wide variety of fields and certain aspects of mentoring for a short project period will be highlighted.

537502 Vanderbiltâ€™s Process Innovation Center: A Multipurpose Facility for Teaching Chemical Engineering Lab and Design Courses

Abstract:

Russell F. Dunn, Ph.D., P.E.

Professor of the Practice

Department of Chemical and Biomolecular Engineering

Vanderbilt University

Nashville, TN 37235-1604

Ph: 615-343-4268

russell.dunn@vanderbilt.edu

G. Kane Jennings, Ph.D.

Department Chair

Department of Chemical and Biomolecular Engineering

Vanderbilt University

Nashville, TN 37212

Ph: 615-322-2707

Kane.g.jennings@vanderbilt.edu

Scott A. Guelcher, Ph.D.

Associate Professor

Department of Chemical and Biomolecular Engineering

Vanderbilt University

Nashville, TN 37212

Ph: 615-322-9097

scott.guelcher@vanderbilt.edu

Vanderbiltâ€™s Process Innovation Center: A Multipurpose Facility for Teaching Chemical Engineering Lab and Design Courses

A state-of-the-art Chemical Engineering **Process Innovation Center** (PIC) has been created over the past seven years at Vanderbilt to allow immersive education of our junior and senior level chemical engineering students. Key core teaching components in this facility has been developed over the past six years. Three key disciplines are emphasized and integrated through the use of this facility for four laboratory and design courses. The key disciplines fundamental to chemical engineering education that are emphasized are:

1. Hands-on operation of lab scale chemical process and product equipment
2. Computer-based design of chemical processes and products
3. Incorporating safety in chemical process and product design, scale-up and operation

A specific description of the integration of chemical engineering lab, design and safety courses will be provided along with a detailed description of the center. This approach will highlight the immersion experience that chemical engineering students at Vanderbilt receive through the use of this center.

536602 Integrating Laboratory Experiments into Chemical Engineering Core Courses

Abstract:

Continuous practice with chemical engineering (ChemE) topics reinforces student learning; and when taught in multiple contexts it is also more likely to lead to true knowledge transfer than simply memorizing information. This process takes time and requires that students see material in a variety of settings over an extended period of time. The senior ChemE laboratory sequence at the University of Kansas has always served to give students a deeper perspective on theory covered in their core chemical engineering classes and recent advancements have been made to improve the integration between the laboratory and preceding fundamental courses.

Three newly commissioned experiments on fluid mechanics, thermodynamics, and chemical kinetics were specifically designed for incorporating into the teaching of their junior-level counterpart courses. Mobile design of these experimental systems allows them to be brought into the classroom where students are able to get a close-up look at how practical experiments are conducted. Projects given to the students using data collected in the senior laboratory also introduce concepts of experimental uncertainty, technical report writing, oral communication, teamwork, and leadership skills prior to ever setting foot in the lab.

This presentation will provide an overview on the three experiments used by students in their first chemical engineering laboratory course and their integration into the corresponding junior-level core courses.

BIOCHEMICAL ENGINEERING LAB COURSE: PREPARING UNDERGRADUATES FOR THE BIOTECHNOLOGY INDUSTRY

Frederick Twigg, UC Berkeley, Berkeley, CA
Shannon Ciston, UC Berkeley, Berkeley, CA
Esayas Kelkile, UC Berkeley, Berkeley, CA
Wenjun Zhang, UC Berkeley, Berkeley, CA

Introduction

As the biotechnology industry continues to grow rapidly, there is a need for chemical engineering graduates who are familiar with modern process equipment and scale-up procedures. A laboratory component to biochemical engineering lecture classes offers chemical engineering students a hands-on environment to apply theory and gain practical experience with biotechnological process equipment which will prepare them as excellent candidates for industry positions. We describe a course newly developed in the Spring 2018 semester to address the industrial need for graduates with practical bioprocessing experience.

According to the National Bureau of Labor Statistics, the annual growth rate for the Chemical Engineering occupation is projected to be 8% from 2016-2026.[1] This is just slightly above the projected national average of 7%. Though this growth rate is moderately fast, it is not as fast as the rapid 10-11% projected growth rate for certain life sciences positions such as biological technicians and biochemists. The technological advancements in the industrial biotechnology fields are major contributors to this growth. The global market for biotechnology is predicted to hit nearly $730 billion by 2025.[2] This is an impressive leap over the $330 billion biotechnology market value estimated in 2015.[3] This presents a unique challenge to the field of chemical engineering, which has a long history at the interface of process engineering and biochemistry.

Unit operations is the canonical keystone course in chemical engineering curricula which combines the understanding of momentum, heat, and mass transfer (and of course chemical reactions) and applies it to industrial applications. The design and operation of these real-world pieces of equipment are intended to prepare students for the many types of processes which are utilized in industrial positions. With over 67,000 life sciences jobs in the San Francisco Bay area, San Francisco is one of the world's major biotechnology hubs.[4] At UC Berkeley, we are right next door to these exciting developments and are working on a new biotech version of unit operations to answer the challenge presented by the need for process engineers with a working

[1] Bureau of Labor Statistics, U.S. Department of Labor, *Occupational Outlook Handbook*, Chemical Engineers, on the internet at https://www.bls.gov/ooh/architecture-and-engineering/chemical-engineers.htm
[2] Grand View Research, August 2017 Report, https://www.grandviewresearch.com/press-release/global-biotechnology-market
[3] Global Market Insights, October 2016 Report, https://globenewswire.com/news-release/2016/10/05/877191/0/en/Biotechnology-Market-size-to-exceed-USD-775-billion-by-2024-Global-Market-Insights-Inc.html
[4] JLL 2017 Report on San Francisco Bay Area Life Sciences Outlook, http://www.us.jll.com/united-states/en-us/Research/US-San-Francisco-Bay-Area-Life-Sciences-Outlook-2017-JLL.pdf?3fc2443a-dd90-47a2-9d90-a75041784d5d

understanding of the trade's equipment. This past Spring semester of 2018, we established these four modules to begin developing what array of units are necessary.

1. Cell growth in shake flasks and cell harvesting
2. Cell disruption, protein recovery, and protein purification
3. Protein activity assays
4. Bioreactors and pilot-scale fermentation

Expanded details on these modules are presented below. Initial feedback from the eight students who enrolled this past Spring for the first offering were positive. One constructive item raised was the perception that some procedures were too codified. The present semester's procedures are testing more flexible variations on the activities to address this.

The scope of this course does not currently include mammalian cell culture or nanobiotechnology. Further additions to this course may include these fields as we grow the facilities and class size. Collaboration between academic units and professional representatives also may help direct the development of this applied laboratory component to better prepare chemical engineering graduates looking at jobs in the biotechnology market.

Overview Of Modules

Cell Growth In Shake Flasks And Cell Harvesting

This unit provides an opportunity to quickly learn basic benchtop techniques and review some of the core synthetic biology topics required to understand bioprocessing unit operations. In designing a bioprocessing course for chemical engineers, it is a challenge to balance time spent covering theory and time spent learning the applied operation and quantitative data analysis necessary for specific units. The course strives to focus on the latter, though students will require sufficient background to follow what is being achieved in a particular unit operation. We have addressed this problem with brief lectures within the lab sessions to review fundamental topics. Laboratory sessions involving growth periods in particular provide downtime were class discussions can be inserted.

This unit covers the preparation of media, basic growth curves with *E. coli*, optical density measurements, choices in bioorthogonal pH buffers, and the use of agar petri plates for bench scale strain isolation and genetic manipulation. Media preparation exposes students to the dichotomies of simple vs complex, defined vs undefined, and minimal vs rich. Media preparation also provides an opportunity for students to explore how salt solubility rules and the Maillard reaction impact which media are more suitable for heat sterilization or filtration (or a combination thereof). After working with a variety of media types, students can apply optical density measurements to demonstrate how media affects changes in growth rates and stationary phase populations. Throughout this unit, towards the end of each lab period, students have the opportunity to practice the most basic molecular biology techniques of plasmid transformations and agar petri dish streaks.

The amount of time spent on this unit is quite flexible and can be tuned to the experience level of the class. If the class has a high degree of prior laboratory experience, sub-topics can be added such as the impact of flask geometry on culture doubling times or the use of computational models to predict the effect of nutrient concentrations on bacterial growth curves. In future iterations of this course, this module should be expanded to include non-bacterial cultures. The growth of *Saccharomyces cerevisiae* and mammalian cell culture are both industrially relevant areas with certain fundamental differences from bacterial cell culture.

Cell Disruption, Protein Recovery, And Protein Purification

The prior module provides ample experiments to collect cell material containing recoverable proteins. The module introduces the plethora of chromatographic approaches to protein purification and emphasizes that industrial purification schemes rarely involve single step approaches. An illustrative concept that jumpstarts the challenges in purification is one of the many available total metabolic maps. It illustrates how cells are complex mixtures containing a few thousand distinct proteins, and an equally diverse array of metabolites. Pulling a singular high purity product from these mixtures is a daunting challenge, and this unit guides students through the engineering solutions that have be refined for industrial application.

In its present formulation, the module covers size exclusion, ion, and affinity chromatographic approaches. The class also includes salt-based precipitation as a non-chromatographic method for crude protein purification. Many of these are done via FPLC, an important unit op in the biotech industry, but manual bench top columns are also demonstrated. A variety of assayable proteins are made in this module, which segues nicely into the following module. In the performance of manual chromatography, students collect samples from each phase of the purification to check via Bradford assay and SDS-PAGE. Analyzing real and supposed SDS-PAGE results can be used to challenge students to predict and troubleshoot possible problems in the expression and purification processes that may have led to the target protein's poor purity (such as poor retention to the column).

The FPLC is a major unit in this course and combines an understanding of the column types with running solvent selection and method development. The FPLC model should at least include common conductivity probe and DAD capabilities for run analysis. If the students have already taken a traditional unit ops course, than the fundamental operation of pumps can be abbreviated, but if students are unfamiliar with the operation of peristaltic and binary pumps, a brief introduction may discuss pump priming, leak identification, the risks of dry running with columns in line, and troubleshooting the source of erratic pressure swings. Students should also have a familiarity with the process flow diagrams used in FPLC software to design flow paths for system flushes, bypasses, and column selection. In current iterations, we used green fluorescent protein (GFP) in the first stages of FPLC operation. The protein was expressed at high enough titers that, during fractionation, students received immediate feedback on how efficiently GFP was loaded and subsequently eluted from the column. This allowed students to optimize and compare both gradient, step, and isocratic approaches to protein elution.

Future developments in this module should focus on student access to method development software. A few chromatography data sets allow students to analyze many system parameters such as dwell volume, theoretical plate counts, selectivity, and retention factors, but the method development itself is a key component to FPLC operation. It would be impractical to provide every student with their own FPLC system, but method simulation could provide every student the opportunity to see theoretical impacts of their choices on protein purification efficiency. The present use of GFP is useful for demonstration but is not a useful protein by itself. It may be beneficial to use fusion proteins, such that the target protein is both visible and directly assayable. Clever use of multiple visible fluorescent proteins prepared in advance of a lab session could also provide opportunities to quickly visualize peak selectivity.

Protein Activity Assays

This module follows directly from the previous module, and students assay the activity of the proteins they have now purified from a number of different chromatographic approaches. This module tends to be among the shorter of the presently taught modules since activity assays can be performed rapidly, and there is less waiting time than in previous modules where cells are growing, or FPLC runs are in progress. The topic currently covers enzyme kinetics (such as with an oxidoreductase that turns over NADPH) and enzyme activity (such as a *Taq* polymerase).

For enzymes kinetics we had students analyze the turnover of NADPH by a FAD-dependent oxidoreductase (the FAD-dependency is convenient for its distinct yellow color when purifying it). The lab demonstrates Michaelis-Menten kinetics as students see the impact of substrate and cofactor concentrations on the rate of NADPH consumption to oxidize an amino acid substrate. We have also used *Taq* DNA polymerase as a substitute for a quality control procedure. Rather than observing substrate conversion, we compare the class-purified *Taq* to a commercial polymerase routinely used in laboratory screening. After trying different extension and annealing conditions, the students can see how their polymerase activity stacks up against commercially available versions.

This module has room to improve and be more representative of industrial quality control procedures after a batch of a protein product is made. As we expand the available equipment and curriculum, this module could possibly be absorbed into the final step of each protein purification as more labs include areas outside the present scope of class.

Bioreactors And Pilot-scale Fermentation

The bread and butter of many industrial bioprocessing operations, this module focuses on the operation of a stainless-steel bioreactor and the control of fermentations. At face value, bioreactors have much in common with the fed-batch reactors that senior chemical engineering students would all be familiar with. The module discusses choices in baffles, agitation rate, impeller speeds, eddying effects, and fed-batch systems for a 15 L pilot scale bioreactor. After a thorough understanding of the bioreactor internals and a safety discussion, we move to clean-in-place (CIP) sterilization procedures and calibration of dissolved oxygen and pH probes.

Once small student teams of about four students have demonstrated safe operation of the instrument, the students can scale up some of the procedures which they have now exercised in the first and second modules. Students can observe some of the massive growth improvements obtainable with high agitation rates and constant oxygenation of liquid media. In addition to comparing protein titers from the bioreactor to flask cultures, students can observe growth curves in both anaerobic and aerobic environments. Exercising process control features as well, the lab takes advantage of *E. coli*'s robust nature and shows how changing pH response PID parameters affects the instrument's ability to react to pH changes in the tank.

In the present development of this course, we have also blended this module with the exploration of small molecule purification. Students grew strains of the soil microbe *Streptomyces*, which are renowned for their biosynthesis of natural products. Though our current lab lacks dedicated HPLC and mass spectrometry equipment, we were able to borrow equipment from UC Berkeley labs to demonstrate basic aspects of organic extraction of natural products from fed-batch grown bacteria. Natural product extracts also provided the opportunity to utilize hydrophobic interaction chromatography, one type not covered in the protein purification module.

This module will benefit from the availability of a wider array of bioreactor types. In addition to fed-batch reactors, students would benefit from operating continuous flow bioreactors. Single-use bioreactor bags are becoming more common in medical bioprocesses (such as antibody production) but may be expensive to implement. Increasing the scope of organisms grown in bioreactors will also help broaden the scope of this course.

Practical Challenges And Future Developments

One challenge of this lab which was immediately apparent in its design was the length of time many bioprocesses require. Despite the rapid doubling time of *E. coli*, a few hours of growth time can be the entirety of a laboratory session. While many steps in bioprocesses are somewhat capable of being frozen and restarted, most procedures have more success if done in continuous execution. While blocks of time that require minimal active wet work can be filled with well-placed lecture material or reagent preparation for the future, it is not always easy to balance each lab period with the active "doing" expected of lab course.

These challenges can be mitigated with well-planned preparation in advance but requires a relatively high amount of personnel time outside of basic laboratory oversight. To make sure that mistakes don't translate into losses of laboratory session time, we often prepared extra versions of protein samples and cell lines outside of class sessions, which could quickly be provided if a student group lost all of their material. Letting students make errors provides an opportunity to learn from empirical observations but having back-up materials can prevent repeating more steps of the procedure than necessary.

As this course develops and expands, the objective is to include additional unit ops mainstream to the current bioprocessing industry. The curriculum should remain flexible to adapt to changes in the field while maintaining the topics grounded in each respective system's fundamental principles. Existing modules can be tweaked as we receive student feedback. From our first eight students who took the course this past Spring, one trend was a desire for more independent discovery. To meet this challenge while maintaining a safe environment and preventing damage to instruments, we have developed less rigid protocols. One example includes challenging four separate teams to make a media with a number of salts and sugars in the minimum number of bottles that can be heat sterilized without causing precipitation or inducing the Maillard reaction with sugars. In the protein purification module, teams are challenged to minimize the number of runs to effectively fractionate GFP and separate it from other contaminants with minimal explanation of purification strategies. As of this writing, the second semester offering of this course is entering the protein purification module. Students seem to have positively responded to the added challenge, but we will see more formal feedback after mid-term evaluations.

Closing Remarks

The assembly of a bioprocessing version of unit operations has been an exciting endeavor, and we look forward to student feedback on the direction of the course to date. The chemical engineering umbrella has included fields in the area of biotechnology for some time now. The field includes the design and manufacture of biologics, agrochemicals, pharmaceuticals, and commodity chemicals among others. It may prove challenging to cover each situation with industrial relevance, but as with canonical unit operation courses, the sampling of equipment and processes should be demonstrative of the types of processes that graduates may experience in the field of bioprocessing.

530929 Case Studies As Vehicles for Enhancing Studentsâ€™ Perceptions of the Broad Relevance of Biochemistry in Chemical Engineering

Abstract:

We have piloted a course that presents chemical engineering case studies at the end of a survey course on biochemistry. All chemical engineering students at Iowa State University are required to take at least one semester of biochemistry. These students bifurcate into two groups. One group has interest in biological applications of chemical engineering and takes a two-semester sequence. The second group has much less interest in biological applications of chemical engineering and takes a one-semester course. Through senior surveys, we measured dissatisfaction with the biochemistry requirement in this second group. This was due in large part to a lack of understanding of the broad impact of biochemistry on chemical engineering industries. Consequently, collaboration between the Department of Chemical and Biological Engineering and the Department of Biochemistry, Biophysics and Molecular Biology was formed to address this problem. We designed a new course containing two components. The first component included a condensed presentation of core biochemistry principles. The second component included chemical engineering case studies that showcased the relevance of biochemistry in the development of industrially relevant products. We will present the design of the case study topics, content delivery approach and assessment. We collected data from end-of-course evaluations, surveys deployed at the beginning and end of the course and senior surveys. Longitudinal analysis indicated that the overall student rating of the course improved. In addition, students who took the course with case studies were more likely to agree that biochemistry is broadly relevant to chemical engineering and disagree that biochemistry is only relevant to healthcare. Future work will be focused towards determining if the case studies result in a better understanding of fundamental topics in biochemistry and motivate students to take additional biology-focused courses.

525314 Student-Driven Process Oriented Guided Inquiry Learning (POGIL) Based Biochemical Engineering Technical Elective Course

Abstract:

This presentation will examine taking an innovative approach to using a student-driven process oriented guided inquiry learning (POGIL) method in a high-level undergraduate Biochemical Engineering technical elective course. The course is intended to assist students in developing advanced critical thinking skills and the ability to apply existing knowledge of chemical engineering concepts in order to solve real-life problems in the food, pharmaceutical, biomedicine, biotechnology and biofuels industries. In this presentation, I will provide details on a) the course components, including student-driven lab activities, capstone design-like individual projects, discussions and activities surrounding process safety and labelling guides for bioprocess industries, as well as b) two-way peer-assessment strategies.

524862 Sensitizing Bacterial Cells to Antibiotics through Dynamic Topographyâ€“Triggered Biofilm Detachment

Abstract:

Microbial biofilm is a leading cause of chronic infections in humans and persistent biocorrosion and biofouling in industries due to extremely high-level tolerance of biofilm cells to antimicrobial agents. In the past three decades, the significance of biofilms has stimulated extensive studies on strategies for biofilm prevention and removal. However, due to the protection of extracellular matrix and slow growth of biofilm cells, controlling mature biofilm remains challenging. Recently, we reported that mature biofilms can be effectively removed (e.g. 99.9% detachment of 48h *Pseudomonas aeruginosa* PAO1 biofilms) using shape memory polymer-based dynamic change in surface topography. Here, we demonstrate that such biofilm detachment also sensitizes biofilm cells to conventional antibiotics. For example, concurrent treatment of 48h *P. aeruginosa* PAO1 biofilms during dynamic biofilm removal increased the susceptibility of released biofilm cells to tobramycin by more than 3 logs (2,479 fold) compared to surface-attached biofilm cells. The observed effects were attributed to the disruption of biofilm structure and increase in cellular activity as evidenced by a 11.8-fold increase in ATP level and 4.1-fold increase in expression of *rrnB* gene in detached cells. These results can guide the design of future combinational therapies for biofilm removal by integrating newly-designed smart dispersion strategies with currently available antibiotics.

536323 The Role of Flagellar Motor Reversals in Swarming in *Escherichia coli*

Abstract:

Swarming in thin fluid films on top of solid surfaces is enabled by the rotation of flagellar filaments in bacteria. To swarm, cells must overcome an increase in viscous drag due to the proximity to the surface. Previous reports suggest that cells likely compensate for the surface drag by promoting surface wetness through an unknown mechanism with the aid of a flagellar switch. The switch helps the filament erratically reverse the direction of rotation between clockwise (CW) and counterclockwise (CCW). The direction of rotation is dependent on the conformation of a switch-component, FliG. Here, we investigated the role of FliG conformation in facilitating bacterial swarming. We observed that swarming was abolished when FliG was locked in the CW conformation ($FliG^{CW}$), but not in the CCW conformation ($FliG^{CCW}$). Swimming was not abolished, so the conformation of FliG is likely critical for swarming. We found that torque generation in $FliG^{CW}$ strains was similar to WT strains. Swimming speed analysis revealed the loss of swarming ability correlated with a reduction in the flagellar thrust in the CW strain, which was likely due to a reduction in the number of filaments per cell. Fluorescence visualization showed that $FliG^{CW}$ swimmers have one less filament per cell on average compared to WT swimmers. Yet, $FliG^{CW}$ swarmers tend to lack filaments entirely (~57% of cells visualized carried no filaments, compared to ~10% in the WT). We hypothesized that there was a link between the function of the switch and degradation of the flagellar assembly; however, TIRF imaging revealed no significant degradation. Although the role of the switch in swarming remains unclear, our assays revealed that WT swarmer cells show a significant decrease in the probability of CW rotation (known as CW_{bias}) suggesting modifications are occurring at the level of the switch, and further pointing to its importance.

528605 Pseudomonas Aeruginosa Single-Cell Level Heterogeneity, Investigated Via Drop-Based Microfluidics

Abstract:

Single cell microbial studies can be challenging using traditional microbiology methods; however, single cell studies are critical in understanding cell-to-cell heterogeneity for interpreting bulk microbial growth behavior. This is especially true in biofilms, or microbial communities that express high levels of population heterogeneity and exceptional resilience due to its heterogeneity. Drop-based microfluidics, or the formation and manipulation of picoliter-volume drops of water-in-oil allows for the encapsulation of single cells, providing a platform for single cell studies. In this work, populations of Pseudomonas aeruginosa are examined at the single cell level within microfluidic drops incubated on a microfluidic chip over a period of 24+ hours. Confocal microscopy tracks the expressed plasmid gfp fluorescence of the cells. The resulting data is used to construct growth curves starting from single cells and to quantify the growth rate and lag phase. Our long-term growth study is performed using a device preparation technique that fills the porous microfluidic device with water-saturated carrier oil and allows for thermodynamic stability between the oil and aqueous phases. In absence of this preparation, drops are observed to shrink, foam, or evaporate from the device over time. Bulk studies of P. aeruginosa cells reintroduced to media after 4 days of starvation shows decreased growth in resuscitating cells. However, single cell studies using this incubation method show that these cells have increased and variable lag phase lengths, rather than decreased specific growth rates (Akiyama, et.al. PNAS, 2017). Our results demonstrate the quantification and monitoring of single cell behavior over 24+ hours, which will allow for a deeper understanding of the role of heterogeneity in bacterial populations.

537121 Differential Response of Mucoid and Non-Mucoid *Pseudomonas AeruginosaÂ* isolates to Interfacial Confinements

Abstract:

Fluid interfaces are energy-rich environments known to influence the self-assembly of small molecules or microparticles, and the formation of thin films by biological entities. The formation of bacterial films at fluid interfaces appears to be a dynamics process, in which cells coping with the existing interfacial energies respond to the entrapment by forming a matrix appropriate to their survival. We hypothesize that cells subjected to interfacial stress adapt and exhibit phenotypic changes essential to their survival. In this study, we are investigating the response to interfacial confinement of two clinical strains of *Pseudomonas aeruginosa* isolated from the airways of cystic fibrosis patients. To evaluate what phenotypic changes at fluid interfaces provide growth advantage, we are comparing a mucoid (PASL) and non-mucoid strain (PANT) of *P. aeruginosa* secreting various level of polysaccharides. We are investigating the mechanical properties of films formed by these cells to understand how they adapt and restructure environments with high interfacial energies. Some intrinsic properties of the cells confer viscoelastic behavior facilitating their adaption as characterized by the apparent moduli of the elastic films observed using pendant drop elastometry and particle tracking. Further characterization of whole genome and the transcriptomes of cells under interfacial confinements will provide new insights on the biological implications of interfacial films.

540115 Bloodmeal-Induced Inhibition of *Plasmodium* infection in Mosquito Vectors Using the Microbial Symbiont *Asaia*

Abstract:

Asaia sp. are gram-negative bacteria that have been shown to colonize the *Anopheles* mosquito midgut. Species of this mosquito are the vectors for malaria transmission throughout the world. Using a paratransgenic strategy in which transgenically modified symbiotic organisms affect their hostâ€™s phenotype, *Asaia* has been engineered to secrete anti-malarial effector molecules into the mosquito midgut, hindering its ability to carry *Plasmodium*. However, the constitutive expression of these effectors causes a fitness disadvantage to the transgenic *Asaia*.Therefore, it is desirable to express these molecules only when *Plasmodium*is present in the mosquito midgut, namely when a mosquito takes an infected blood-meal. Putative conditional promoters have been cloned into the plasmid pGLR1, which has a promoterless GFP reporter. These plasmids were then transformed into the lab strain *Asaia sp.SF2.1* and plated on minimal media enriched with blood. GFP fluorescent colonies were collected and screened both in-vitro, in liquid and solid media, as well as in-vivo, inside the mosquito midgut, to isolate the promoters that were only induced when blood meal conditions were present. Once the conditionality of these promoters was validated, they were cloned into a plasmid secreting the anti-malarial effector scorpine and transformed into *Asaia.* The increased fitness of these conditional strains was validated against a constitutively expressing control before testing each strain for *Plasmodium* inhibition. The anti-malarial strains were cultured in mosquitoes that were then fed on *Plasmodium*infected mice. The conditionally active promoter strains showed a significant reduction from the wild-type bacteria in the number of oocysts the pathogen formed as well as a significant reduction in disease prevalence effectively inhibiting the mosquitoesâ€™ ability to transmit malaria.

537630 Dynamics of Biofilm Elimination on Thermally Shocked Biomedical Surfaces

Abstract:

When bacteria colonize a medical implant surface, they form a biofilm which cannot be eradicated chemically. The current standard of care is surgical explantation of the device and surrounding tissue, with eventual reimplantation of a replacement device with twice the probability of infection. These infections are a $5 billion problem in the U.S. alone, impacting over 100,000 patients annually. One approach to mitigating these biofilms is *in situ* thermal eradication by generating a thermal shock directly at the device/biofilm interface. We have developed magnetic nanoparticle / polymer composite coatings which heat rapidly when exposed to an externally applied alternating magnetic field, and have quantified immediate bacterial population reductions of up to 6 orders of magnitude (i.e., "6-log" reductions).

Moreover, under certain circumstances the biofilm bacterial population continues to decrease for hours after the thermal shock is removed and physiological temperature is restored, resulting in complete eradication of the biofilm. Re-incubation studies at a variety of shock temperatures (50 – 80 °C) and exposure times (1 – 30 min) showed that thermal shocks producing immediate population reductions greater than 4-log resulted in complete elimination of the biofilm approximately 4 hours later, regardless of the temperature/exposure time combination used to achieve the immediate reduction. Biofilms subjected to thermal shocks with more modest immediate population reductions (less than 4-log) grew back to their initial population density within 24 hours. These investigations were extended to biofilms with different culture conditions producing pre-shock population densities 100 times greater than the earlier biofilms (nearly 10^9 CFU/cm^2). Despite the substantially larger initial population density, the same immediate population reduction (4-log) was required to prompt eventual die-off of the biofilm, even though the population density immediately after the thermal shock was 100 times greater than in the previous studies. This suggests that the degree of thermal shock required to eradicate a biofilm is independent of the biofilm's initial population density.

Furthermore, this approach may be integrated with other modifications to the biofilm/device interface to prevent or eliminate biofilms with even milder thermal shocks. Investigations of combined thermal shock and antibiotic exposure yielded population reductions whose orders of magnitude were greater than the sum of the orders of magnitude reduction for either treatment alone, enabling significantly milder thermal shocks to provide the same efficacy and providing an impetus for localized delivery at the device interface of antibiotics which are by themselves insufficient for biofilm elimination.

535582 Modelling Microbial Microenvironments through Encapsulation of Synthetic Communities

Abstract:

In the field of microbiology, it is believed that less than 1% of microbes are presently known and have been studied. The inability to culture many microbes using traditional methods, such as plating or suspension, has led to a vast gap in our understanding of microbiomes. To overcome this challenge, we have developed a novel microfluidics system which enables us to encapsulate microbes in a semipermeable membrane on the scale of nanolitres, eliminating resource competition and effectively increasing cell communication of isolated microcolonies.

In this study, we are introducing a mock community developed in the nanocultures with five microbial species, including *Pseudomonas aeruginosa, Escherichia coli Nissle 1917, Staphylococcus aureus, Streptococcus mutans and Candida albicans.* We determine how population density differs when cells are cultured in nanocultures as opposed to culturing them in suspension in a flask. After an incubation period of 24 hours, the nanocultures are ruptured and DNA is isolated and sequenced from viable cells. 16srRNA sequencing allows us to determine the abundance of each species in both the nanocultures and flask cultures, thus we can determine how the species interact with each other under the constraints of each system. We anticipate that our microfluidics device will be able to capture unculturable microbes in future applications, from a variety of different environments, such as oral, gut, marine and soil, allowing for unknown microbe discovery, and furthermore, a more accessible and in-depth study of these unknown microbes.

523853 Getting Active: How the Asee Summer School Shaped My Teaching Practices

Abstract:

Understanding how current students learn and engage in a classroom is a topic in which not many young professors are well versed. Traditional educational techniques, such as a straight lectures or a PowerPoint "slide show", are so prevalent that many new faculty or faculty candidates have very few examples from their own time as students to help shape their future teaching practices. Fortunately, there are a number of passionate educators among members of AIChE and the American Society for Engineering Education (ASEE) community who have identified this problem and organize many resources and events for junior faculty to develop this skillset. Opportunities such as ASEE's Summer School for Chemical Engineering Faculty, helped expose new faculty, such as myself, to ideas like active learning, time management, and the outside resources available to chemical engineering educators.

To engage students more during lecture, I developed course content based on active learning. Lectures were designed so students solved problems or derived key formulas in teams during frequent in-class activities, rather than providing the information in a traditional lecture. Students were given the opportunity to interact with each other or myself as they completed the activity, before a "random name generator" selected one or more students to offer an answer. Furthermore, courses were structured around the concept of "handouts with holes". Instead of creating pre-made work sheets or a PowerPoint "slide show", handouts missing key information or ones that need to be completed during class were provided. This method was employed by preparing very stripped down lecture notes in PowerPoint that were subsequently uploaded for the students to print prior to class. As class proceeded, students filled out the notes, which shifted the responsibility of note taking to them, while still keeping the day's lecture more organized.

To better balance my own responsibilities, I adopted various new time management techniques focused on grading and content creation. One method was the number of student's work graded for each assignment. Homework was done in groups, but each student was responsible for submitting their own hand-written copy. To use time more effectively, only one assignment from each group was randomly selected to be grade. Once graded, each group member received the same grade. This allowed for more in-depth assignments due to the reduced grading time burden. Another method was creating answer sheets for the students to use during exams. Making a stream-lined answer sheet for each student to use helped reduce the time and effort it took to grade the assignment. Lastly, resources like the AIChE Concept Warehouse helped build content and refocus the fundamental way I created exam problems for the students.

The ASEE Summer School experience was seminal in changing the way I view education. The practices, workshops, and talks the junior faculty were given access to this early in their career helped evolve my teaching style completely. Furthermore, the program offered the chance to make contacts with other young faculty, all of whom will be my colleagues as we start our journey in academia together. The exposure to these new concepts and colleagues will continue to shape my future academic career.

538842 Incorporating Teaching Methods from the ASEE Summer School during My First Year on the Job

Abstract:

I attended the ASEE Chemical Engineering Summer School before my first semester as faculty at the University of Texas at Austin. In Fall 2017, I taught Chemical Reactor Design and Analysis to a class of 70 students. Several methods from the summer school were incorporated into my teaching, including the use of â€œnotes with gaps.â€ Based on course survey data, 74% of students strongly agreed that the course materials were helpful to them, compared to 43% of students from previous semesters. In Spring 2018, I taught the same course again to a class of 80 students. During this semester, I have also incorporated more interaction into my lectures by using conceptual questions from LearnChemE and the AIChE Concept Warehouse. During the middle of lecture, students are given a chance to discuss the questions before being polled for their answers. In this talk, student feedback from both the Fall 2017 and Fall 2018 semesters will be discussed.

532831 Combining the Luck of the Draw with Individual Accountability in Team-Based Learning

Abstract:

Group work at any level in lecture-based or laboratory courses is a challenging task for the students and the instructor. Inspired by several workshops of the ASEE 2017 ChE summer school such as game-based learning, active learning, and groups and conflicts, the luck of the draw was implemented to promote individual accountability in a team-based project. This approach was used in an undergraduate level Transport Phenomena class in which the students selected a topic of research based on chance. Depending on the number of students and groups, each team picked a Transport Phenomena subject area randomly. The raffled areas included Momentum, Heat or Mass Transport. After drawing a topic, the students were assigned an open-ended project on which each group developed a problem statement with an innovative solution that involved the raffled topic area. Once the problem statement and the solution were turned-in to the instructor, a brief presentation was made to the class, but the rules stated that only one of the students from each group was responsible for this task. Again, by the luck of the draw, the presenter was selected just minutes before the presentation. After implementing this method, it is expected that all students should be accountable and engaged when working in a group and in the project. Surveys of individual performance for each group were collected to obtain a better understanding of the group dynamics through this technique. In addition, all the presentations were graded by every student in the class. This method has been applied to self-selected groups, however, to validate it, future studies should focus on randomly selected groups or instructor assigned teams.

530859 Improvements in Teaching Core Undergraduate Courses after Chemical Engineering Summer School 2017

Abstract:

After attending Chemical Engineering Summer School in 2017, I made some significant improvements in my Fluid Mechanics course at The University of Utah. Around 90 students were enrolled in Fall 2017. This was my second time teaching this course; the first time was in Fall 2016. My course evaluations for 2017 versus 2016 in Fluid Mechanics were significantly improved, largely due to modifications made through ideas obtained at the Summer School. One of the major additions to the Fluid Mechanics course was to add a video project, where the students were tasked with producing a short video to explain a fluid mechanics topic to a general audience. The students were exposed to similar videos throughout the course on â€œYouTube Fridaysâ€ in order for them gain ideas for how to structure the videos. The videos were of overall very good quality and clips of these videos will be presented. Other changes to the course will be presented which include an emphasis on daily class objectives and examination study guides.

Additionally, in Spring 2018, I began teaching the undergraduate Chemical Reaction Engineering course. This course was switched from a 50 minute course three times per week to an 80 minute course twice per week. This gave me more flexibility to try some new teaching methods. Ideas from the Summer School such as incorporating active learning were employed in this course and examples of active learning activities that were used will be presented.

533414 How Summer School Improved Our Safety Education in the Unit Operations Laboratory

Abstract:

AICHE: Professional Development Committee Liaison: How Summer School Improved My Teaching - Oral Session

Tracy Carter, Samira Azarin, Janie Brennan, Elizabeth Hill, Amy Karlsson

How Summer School Improved Our Safety Education in the Unit Operations Laboratory

Industry has identified a skills gap between what is needed and what students are able to do by the time they graduate.[1] Creating change in knowledge, skills and attributes of students to meet industry needs in the undergraduate curriculum can be difficult. However a change model proposed by Henerson et al[2] has shown that effective changes can be made. The change model consists of four aspects: 1. *Developing a shared vision*, 2. *Enacting policy*, 3. *Disseminating curriculum and pedagogy*, and 4. *Developing reflective teachers*.

The ASEE summer school created the opportunity for developing a *shared vision*. Five unit operations instructors from five different universities across the country had the occasion to discuss the opportunities and challenges they face teaching this core chemical engineering laboratory course. One issue that resonated with all instructors was safety. This is because safety and safety education is driven not only by *ABET policy*, but by the core ethics of engineering.[3,4]

Over the past year these instructors collaborated on identifying the safety goals and outcomes for their respective courses and have identified outcomes that are similar between the various institutions. They have worked together to create a change in undergraduate safety education, specifically in the area of identifying major regulations. This outcome has been identified by AICHEâ€™s Safety and Chemical Engineering Education (SAChE) Committee as a key learning outcome that meets ABET criteria.[5] The next year will be focused on disseminating curriculum that meets this outcome and reflecting as a group how well the outcome was met.

1. Luo, Y., Westmoreland, P. R., Alkaya, D.,Alves da Cruz, R. V., Grossmann, I. E., Provine, W. D., Silverstein, D. L., Steininger II, R. J., Talbot, J. B., Varma, A., McCreight, T., Chin, K, Schuster, D., *Academia-Industry Alignment: Expectations about New Graduates, an NSF-Sponsored Study led by the American Institute of Chemical Engineers*. 2015.
2. Henderson, C., A. Beach, and N. Finkelstein, *Facilitating change in undergraduate STEM instructional practices: An analytic review of the literature.* Journal of Research in Science Teaching, 2011. **48**(8): p. 952-984.
3. *Criteria for Accrediting Engineering Programs, 2016 â€" 2017*.[cited 2017; Available from: http://www.abet.org/accreditation/accreditation-criteria/criteria-for-accrediting-engineering-programs-2016-2017/#program].
4. National Society of Professional Engineers, *Code of Ethics*.[cited 2018; Available from: https://www.nspe.org/resources/ethics/code-ethics].
5. SAChE Recommendations for ABET Safety Content in Chemical Engineering February 18, 2010. [cited 2018; Available from: http://sache.org/SACHEGuidelinesForABET.pdf].

538558 Opportunities and Obstacles: Using Board Games to Engage Students in Deeper Analysis of Societal Issues with (Potential) Genetic Engineering Solutions

Abstract:

Inspired by the Summer School session on using games in teaching, I have designed and implemented a board game project for a brand new genetic engineering elective I am currently teaching. For the semester project, teams of two to three students investigated a societal issue that has potential genetic engineering solutions. Teams had to not only identify the specific issue they were focusing on, but also choose a time period and geographic location for the game. In the first half of the semester, the teams identified stakeholders, analyzed the social, political, economic, historical, and religious influences on their issue in the context of its time and place, as well as explored the potential genetic engineering technologies that could offer solutions. In the latter portion of the semester, the student teamâ€™s designed educational, cooperative-play, board games for an audience of their peers. Each team had to identify three learning objectives, as well as develop a pre- and post-play assessment to evaluate the effectiveness of their game as an educational tool. The game basics (design of board, structural elements, and baseline rules) were provided by the instructor to better ensure success of the final products and create a manageable project. Full details on the project scaffolding, as well as the final outcomes of testing the games through play, will be presented.

531022 Flipping the Classroom to Increase Student Engagement in an Elective Biochemical Engineering Course

Abstract:

This presentation will focus on how tips and techniques from the ASEE Chemical Engineering Faculty Summer School were incorporated into an upper level biochemical engineering elective course to increase student engagement and interest in the topic. The author has taught this course five times and has made efforts with each offering to modify the course based on student input. In Louisiana, the majority of students enroll in chemical engineering based on their exposure to an interest in working for the petrochemical industry. As a result, many students are unaware that chemical engineers can work (or perform research) in the fields of health care and pharmaceuticals that involve learning about biology, biochemistry, and biochemical/biomedical engineering. Thus, the majority of students who enroll in the elective course may be interested in learning something about the topic but donâ€™t want to put too much effort into the course due to their enrollment in other core chemical engineering classes. Additionally, the studentsâ€™ preference is to work problems and examples in class instead of listening to traditional lectures. This is challenging in a biochemical engineering course because there is a substantial amount of biological background required. After the Summer School, the author decided to implement a flipped classroom approach to provide the students with greater opportunities for active learning during class time. To provide sufficient background on the biological content, the author elected to use in-class Kahoot! quizzes to engage and assess the students on their out-of-class reading assignments. This provided students with a chance to learn content on their own, creating an interactive in-class environment as opposed to traditional lectures. To teach the students about solving problems related to enzyme kinetics and cell growth dynamics, the author developed videos for the students to watch outside of class deriving the equations and balances used to solve problems. These videos allowed the students to work problems in class to solidify their understanding of course content. In this presentation the author will provide insight onto how these new instructional methods were incorporated into the class and discuss the informal assessment of student performance, course evaluations, and feedback on the effectiveness of flipping the classroom. Additionally, the author will provide feedback on how to implement a flipped class in a larger (>60 student) classroom setting.

525995 The Startup Method to Managing Large Classes â€“ a Technique Inspired By Asee Summer School 2017

Abstract:

Managing an engineering course in the 50 to 100 student size can be difficult as students still expect a high level of one on one interaction with professors. To meet this expectation, I recently implemented a new classroom management practice for a 70 student section of Chemical Engineering Numerical Methods that blended elements learned from ASEE Summer School in 2017. Prior to starting class, the students are surveyed to determine their post baccalaureate plans and were grouped into ten startup teams according to their career or research interests (advanced materials, agriculture, consumer products, downstream petroleum, energy, food technology, law and business, medical devices, pharmaceuticals, and upstream petroleum). Slack was used as our team messaging and collaboration software, were each startup had a private channel and the class had a public channel. On the first day each student uploaded a picture and content to their slack profile. This tool proved to be invaluable to manage the amount of messaging (>4700 peer to peer and peer to instructor posts through the semester), file sharing, and memorizing student names. Examples will be shown of peer to peer instruction that occurred using this platform. Startup teams sat together in class, worked on class problems, submitted group problem as part of their weekly problem sets, and developed an independent term project that demonstrated numerical methods on real world data. Examples of these term projects will be shared as part of this talk. The outcome of this new course management was overwhelmingly positive. Many more students were engaged than in previous sections and retention of core concepts seems to have improved.

536891 Experiences of Embedding Safety throughout a Chemical Engineering Program

Abstract:

The Department of Chemical Engineering at University College London has recently gone through a restructuring of its undergraduate programs and are now providing an Integrated Engineering Program (IEP) which has a strong focus on inter-disciplinary skills, research-led teaching and project-based learning. At the heart of the program still lies the key fundamentals: a thorough understanding of engineering fundamentals, a strong focus on team working and professional skills, and above all, a way of thinking and a culture that considers safety and sustainability as the core of every activity. How this safety learning is imparted to engineering students in an academic environment can be a challenge for educators, particularly when the activities are to be integrated and coordinated across an entire program. We will describe our experiences of implementing such a program-wide focus, and reflect on the first four years of its implementation.

537403 Preparing Students for Design Experiences in a Global Setting

Abstract:

International educational opportunities for engineering students include, but are not limited to, internships, study abroad programs, and humanitarian programs such as Engineers Without Borders (EWB). Since 2007, a two-semester Global Engineering Outreach (GEO) course in the College of Engineering and Technology at ____________ University has enabled engineering and technology students from multiple disciplines to design and implement humanitarian-based engineering projects. Energy, water, sanitation, and health projects have been implemented in Tonga, Ghana, and Peru. Currently, the projects focus on the needs of communities in Peru.

The question arises as to whether students can adequately design and implement projects without traveling to the country prior to implementation. Each year in the GEO course, a new group of students are divided into four or five teams to design projects selected by two Peruvian communities. Following the two semesters of project development, an implementation trip occurs in May. The GEO class has interacted with these communities for several years so strong relationships exist between each community and the GEO faculty. To facilitate cultural and social preparation for students in the upcoming class, students during the previous yearâ€™s class take videos and obtain community information during the implementation trip to share with the upcoming class members. Significant efforts are also extended within the GEO course curriculum to 1) receive training on social impact assessment and communication, 2) review past reports and videos to increase understanding of the culture, 3) interact on a regular basis with the community throughout the design process, 4) implement social constraints along with technical constraints in the design process, and 5) document all social interactions with the community. Assessment occurs during December, April, and June to assess the effectiveness of cultural preparation prior to and following the implementation trip.

The curriculum methods and associated assessment on cultural preparation and training will be presented to help address the question as to whether students can be adequately prepared for designing and implementing projects without traveling to the country prior to implementation. A brief discussion of future plans for strengthening the cultural preparation will also be presented.

532188 Research Symposium for Engaging Students in Undergraduate Research

Abstract:

Research is always in the need of innovative and fresh ideas; that is why the engagement of students early during their formation is an effective way of instilling in them skills and structured thinking that can be helpful not only for the researchers but for their formation. Undergraduate research has been shown to have strong benefits in terms of how the students develop in the professional fields and how these experiences affect the matriculation in PhD programs [1-4].

Students generally have access to formal research experience during their undergraduate formation by collaborating with professors in their laboratories during their spare time or in summer programs, and in most of the times, the first approach is always time demanding not only for the students but for the different elements of the research group (graduate students, postdoctoral associates, research associates and even professors). In our experience, we have detected three main deficiencies when new undergraduate students collaborate with the research groups:

1. The ability to search, sieve, understand and structure information from peer reviewed journals or other sources of information.
2. Express and defend their ideas in a structured way with scientific fundaments.
3. Structure a valid methodology to begin performing theoretical or experimental activities.

For that reason, we decided to implement different activities that will help solve those deficiencies during the student formation before they get involved with a research group. Our goal was to reach students during the first half of their formation. Two strategies were followed: the development of projects during Integration Laboratories and the participation in a Research Symposium. The first strategy has been applied during the last two years and still we are analyzing the results of the improvement of the projects. The second strategy has been applied for 6 consecutive annual editions and we have observed improvement in the participation of students in research since the first edition.

The Symposium is named "Symposium on Surface Chemistry, Colloids and Materials". The event is organized for students in the first semester of the third year as part of the graded activities of their course in Surface Phenomena. We decided it would be the best course due to the diversity of the field, and that students already have acquired knowledge in organic, inorganic, analytical chemistry and unit processes.

The Symposium organization takes place along the course with different deliverable and graded activities:

1. Month 1: Selection of an article of interest in databases. One session is dedicated to explaining how to access the journal databases, types of publications, contents of an article. The session is given with the help of the library and database managers. Students must select one article of their interest, in line with the topics of the course and no older than 5 years. The teacher validates the selection and gives recommendations in terms of difficulty of understanding.
2. Month 2: Generation of an abstract of the article. The library personnel are responsible of providing the full articles to the students, which they must read, analyze and generate a new abstract with their own words. In this stage, the teacher is responsible of guiding the understanding of the article, revising the redaction and not avoiding lack of any kind of reference including the article itself. With this material, the teacher generates the program and first memories of the Symposium.
3. Month 3-4: Preparation of the presentations. During this stage, the students prepare their presentations either in oral or poster format depending on an aleatory distribution of the works. The students receive a guideline in terms of the format and length of the presentations. Training sessions are also organized to prepare students for the final presentation.
4. Symposium Day During last month: The format of the symposium has evolved along the years; however, the main intention is the same. There is an open invitation to the symposium, thus authorities, teachers, professors, fellow students and family assist. The students present and defend the work in the selected article either in oral or poster format.

The activity has caused different reactions along the years: During the first session, only students assisted, they presented either in English or their native language (Spanish) so they could feel more comfortable during the session. In the following editions, the official language became the English and family and fellow students began to get interested in the event. During the last two editions, fellow professors and graduate students from different institutions agreed to assist the symposium and grade students, in some cases even webcast was used to communicate with professors in institutions as Stanford University and University of British Columbia. For the following session, the Symposium is meant to have invited talks and presentation of the work done at the research groups of the University.

In terms of quantitative evaluation of the symposium, the number of students participating with the "Process Engineering and New Materials Research Group" of the Chemical Engineering major at La Salle University Mexico has increased notoriously becoming the second research group with more student participation in the whole University. Students also comment that the experience was interesting in terms of trying to understand specialized information, presenting and defending their work in a non-native language. The event has become part of the signature activities of the Chemistry Sciences School and has gained recognition from the dean of the University.

The Research Symposium has been a success in terms of formation of students. Not all students will focus on research, however, the experience gives them formation in structuring ideas, manipulating information and expressing ideas that will be useful in any professional activity they develop.

[1] Russell, S.H.; Hancock, M.P.; McCullough *Science Education Forum,* **2007,** *316,* 548-549 "Benefits of undergraduate research experiences"

[2] Dea, P.K. *J. Chem. Ed.* **2000,** *77(4),* 432 "CUR 2000: The many facets of undergraduate research"

[3] Mabrouk, P.A. *J. Chem. Ed.* **2009,** *86(11),* 1335-1340 "Survey study investigating the significance of conference participation to undergraduate research students"

[4] Mohrig, J.R.; Wubbels, G.G. *J. Chem. Ed.* **1984,** *61(6),* 507-508 "Undergraduate research as chemical education"

522037 The Formation of the Dakar American University of Science and Technology in Senegal

Abstract:

A critical weakness of formal education is relating theory to practice. In 2016, we at Engineering Research Consulting were invited to help start a STEM-focused university in Senegal that would address the need of a practical education. The university is called Dakar American University of Science and Technology (DAUST). It is located in Somone, southeast of the capital Dakar. DAUST opened its doors in January of 2017.

To address the issue of practical teaching, heavy emphasis is placed on seminars from working professionals, field trips, internships, international exchanges, and lab work. Professors with industry experience are highly valued at DAUST. Although the official language of Senegal is French, classes are taught in English to prepare students to share ideas with the rest of the scientific community.

DAUST offers science and engineering courses, including chemical engineering, at the bachelorâ€™s and graduate levels. DAUST has several research centers including the Center for Energy Research and Sustainability (CERS). Research topics at CERS will center mostly around energy (both fossil and renewable) and water.

539115 Smart Materials and Microfluidics Research Practicum for Chemical Engineering Students: A Case Study

Abstract:

Microfluidics is an interdisciplinary area combining chemistry, biology, fluid mechanics, nanotechnology and engineering. It deals with liquid flows in channels with the diameter between several micrometers and a millimeter, reduces dramatically the amount of required reagents; only minutes are needed to carry out complex reactions and analytic procedures which usually require weeks for preparation. In addition, microfluidics provides unique conditions for synthesizing and fine-tuning nanoparticles, as well as making drug delivery complexes which cannot be obtained in bulk solutions.

Microfluidics is a tool being intensively studied and used in the USA, Europe, Japan and China. The analysis of selected publications by Andrew de Mello (Imperial College London), Steven Wereley (Purdue University), Anna Shelley (Carnegie Mellon University), Olga Vinogradova (Moscow State University), Takasi Nisisako (Tokyo Institute of Technology), and Jean Christophe Baret (Max Planck Institute in Gottingen, Germany) reveal a sustainable interest of the world scientific community to fabrication of microfluidic chips, exploring their analytic potential, adaptation of research methods to a microscale, and synthesis of new nanosystems in such chips. In other words, microfluidics is expected to revolutionize chemistry as microelectronics did it for electrical engineering.

MOTIVATION

Polymers and surfactants are indispensable components of living matter and chemical industry processes: nucleic acids and proteins, detergents, enhanced oil recovery reagents, and drug delivery systems. They can easily form nanoparticles by self-organization. They are 'smart systems' as they can be easily controlled by many factors, such as pH, temperature, solvent, and etc. Thus, it is not a surprise that they are common objects of soft matter nanotechnology.

It is thrilling that practical application of such systems often involves microchannels (!): blood capillaries, micropores of catalyst carriers, 3D-printing systems, and etc. If we want to know how to apply them properly, we have to study them in microchannels. Microfluidics, however, mostly deals with flow hydrodynamics, fabrication of chips and chemical analysis. Polymer-surfactant systems are also mostly studied in bulk solutions, and not in microchannels. Several publications discussing polymers and surfactants in microfluidic devices are rather individual papers than a focused approach. Thus, a systematic and focused study of polymer-surfactant association in microfluidic environment will be an excitingly new interdisciplinary area which will contribute both to fundamental knowledge and practical applications of such systems.

THE PRIMARY GOAL OF THIS PROJECT is to synthesize and characterize polyelectrolyte-surfactant nanocomplexes and to tune their properties with the use of microfluidic approach.

The secondary goal was to develop a study module consisting of several courses in English involving microfluidics and smart materials to train visiting students at Kazan National Research Technological University, Russia

Intermediate goal: to attract grant support in science and education to support research and academic aspects of project implementation.

Preliminary project implementation results:

The project was already supported two grants: the local (Russian) academic mobility grant to do research at Carnegie Mellon University in the area of microfluidics.

The second grant was given by the federal foundation to support development of the stidy program in the area of smart materials and microfluidics

Anticipated project outcomes with the attracted grant support:

I am confident that this project will shape my scientific career for the next 4-5 years. First of all, I will use my project experience to establish a microfluidic experimental set-up at the Department of Physical and Colloidal Chemistry and continue my research in microfluidics involving dynamic light scattering, microscopy and tensiometry methods. I will disseminate the project content in a series of research and methodological publications. Therefore, a new vibrant research area will be launched in my university to make a fresh contribution to the Russian science (where microfluidics is not represented well).

This research project will generate new knowledge in the US, Russian and world science. I hope that obtained quantitative experimental results on the sizes of nanocomplexes, their phase behavior in microfluidic channels, the use of microfluidic chips for fast synthesis of such complexes, and tuning their properties will contribute to understanding the behavior of proteins and nucleic acids in capillary microchannels and making more efficient drug delivery systems.

As I am both a scientist and an experienced internationalization administrator, I will strengthen the link between Carnegie Mellon University and my home university in Russia. Specific projects in research are the best 'drivers' of international partnership: they always involve students, faculty, and grants. I will apply for the local grant program to invite the faculty of Carnegie Mellon University to my university to give lectures and workshops in microfluidics. This project will also allow me to approach the administration of my university to apply for the Megagrant of the Russian Federal Government to invite the distinguished faculty of Carnegie Mellon University to create a new laboratory in my university.

Finally, this project will form create a new research and academic environment for chemical engineers at Kazan National Research Technological University who are eager two enrish their expertise with deep understanding of smart materials and microfluidics.

516784 Teaching Pharmaceutical cGMP Concepts Using 3DÂ Manufacturing Plant

Abstract:

In the context of education for chemical engineering, the learning is always made more interesting and appealing by providing real-life examples, coupled with problem- or project-based learning. Recent development in computer-based learning (e.g. virtual process plant) has been well received, and has aided learning in courses related to control and unit operations. These teaching pedagogies have proven their efficacy in training chemical engineering graduates with high cognitive and problem solving skills. However, there remains a gap in the training for industry-ready graduates: students should be given more opportunities to develop skills to solve every-day industrial problems in a cost-effective manner. The ideal solution to bridge the gap would be a fully functioning industrial manufacturing plant, where students have full access to a variety of learning activities. But this is rarely achievable given the constraints on cost, time and space.

An alternative is to use model of the manufacturing plant as the basis for skill training. This study aimed to explore the use of innovative teaching tools to enhance the applied industrial component in the cGMP (continuous Good Manufacturing Practices) course of SITâ€™s pharmaceutical engineering program. In the cGMP undergraduate class, the students are employed as employees in the 3D pharmaceutical manufacturing plant. They are tasked to carry out cGMP related activities, using scenarios generated from the model plant. The feedback from the students are generally positive. The model plant is a great tool for visualization, which is especially helpful for students without any industrial experience. The development and implementation of the proposed teaching pedagogy have provided a fresh look into the practicability of conventional chemical engineering education, and also add value to produce industry-ready graduates.

529271 Global Human Engineering Projects and Initiatives to Enhance Student Learning and Strengthen the Curriculum and Program Accreditation Efforts

Abstract:

Human engineering projects can be an important vehicle to strengthen student's technical skills while enhancing their understanding and commitment to the responsibilities of the chemical engineering profession to society. This work focuses on enhancing the curriculum and student learning while meeting ABET requirements through global humanitarian engineering projects. How to best design and implement these projects will be discussed in detail and integration into the classroom will be highlighted. In addition, the work of the Global Societal Impact Committee (GSIC) will be discussed in the context of how AIChE can and does contribute to inclusion of global human engineering in the classroom and the profession. The GSIC was established in 2014 as part of the reorganization of the AIChE Societal Impact Operating Council. The Committee was structured to support global initiatives consistent with the Institute's objectives as the ***Global Home of Chemical Engineers*** and supports students' humanitarian engineering work through grants. GSIC members also present at AIChE student regional conferences and give workshops with students to raise awareness and encourage students' interest and involvement in global humanitarian engineering. Humanitarian engineering projects cover ABET's Criterion 3 Outcomes, especially the new Outcomes 2 – 5 and 7. AIChE collaborations with Engineers Without Borders – USA (EWB-USA) can effectively address ABET Criterion 3 Outcomes and integrate global humanitarian engineering in the chemical engineering curriculum. These collaborations and integration in the chemical engineering curriculum not only serve to comply with ABET requirements, but they strengthen the curriculum and are a significant benefit to students.

523648 An Example of How the Scientific Research Provide New Material in the Teaching of Transport Phenomena

Abstract:

Scientific research in the engineering area, normaly, is carried out independently to the university teaching. This means that the scientist achieves new scientific knowledge and they are reported in scientific papers, divulgated in conferences or congresses, but rarely the new procedures implied or deducted in the scientific research are taken to the classrooms or to the undergraduate laboratories

This means that the priority of the researcher is not to use his findings to reinforce the teaching, but his research line.

In this material, two examples are provided how the results derived from a scientific work are going to pose a didactic problem that can be used in the transport phenomena classes, to illustrate a mass transfer problem.

The research work is related with hydrogen production through water dissociation using heterogeneous photocatalysis. This is a very important topic due the oil is going to be depleted and it is mandatory to look for alternative energies, such as solar, eolic, nuclear, geothermal, biomass, hydrogen, etc. Hydrogen is an energy vector, with a very high calorific value and its combustion does not generate CO_2. It can be produced using the methods: Electrochemical, biomass, catalytic reforming of hydrocarbons, photocatalysis, etc. This last one is a big promise due it has the potential of using the solar light, water, and zero ambient contamination.

The experiments were carried out in a reactor so called Photo – CREC- Water II- Modified, using water, ethanol as electron scavenger and novel catalysts made with titanium dioxide doped with palladium (0.5, 1, 2%). The samples taken from the reactor were analyzed in a gas chromatograph, and the obtained compounds were hydrogen, CO_2, CO and methane.

Heterogeneous Photocatalysis is performed illuminating the titanium dioxide with UV-A light (365 nm), to excite the semiconductor and to promote an electron from the valence to the conduction band. In this way, a couple of electrical charges are generated, an electron and a positive hole. The electron is going to reduce other species such as oxygen or a proton. The hole will oxidize the negative ion OH to convert it in a free radical OH, which will react with the scavenger (ethanol). In this form, reduction- oxidation reactions will be carried out.

Water is dissociated in the proton and the negative radical OH^- due the natural value of pH. To produce hydrogen, the proton reacts with the generated electron and the OH^- with the positive hole, producing a free radical H°. Then, two free radicals H° associate in a covalent bond to produce molecular hydrogen. By the other side, the negative radical OH^- is oxidized by the positive hole, generating a free radical OH° which will oxidize the methanol to eventually produce water and CO_2, the total oxidation.

It is necessary to avoid the presence of air or oxygen in the system, due the oxygen will compete with the proton for the electrons, and we want to favor the proton reduction only.

Experimental results indicate that hydrogen profiles with time have a linear behavior for all the used catalysts, and for TiO_2 doped with 2%Pd, 1.2 micromoles are obtained in 6 hours of reaction, which

indicates the photocatalysts can produce hydrogen and now it is important to improve the technology.

Now the challenge is to predict the transport of the protons from the liquid phase to the surface of the catalyst, and to determine the molar fraction of the protons in the boundary layer surrounding the catalytic particle, to determine the expression for the molar flux of the protons arriving to the limit of the boundary layer and then that for the produced molecular hydrogen, its transport in the liquid to the liquid-gas interphase, and then its transport in the gas phase where a sample is collected by the syringe needle. Also, it is important to determine the pH value at which the reaction rate is controlling and the transport of the protons is fast.

Problem one.

The algebraic development that follows can be incorporated as a pedagogic example that could be taught in classroom, or incorporated as examples or proposed in the section of problems to solve by the student.

A mass balance can be established: The transfer of protons from the liquid to the surface are equal to the reaction rate of the electrons with the protons plus the accumulation of protons, and the equations are written, achieving a ratio between a dimensionless number containing the intrinsic kinetic rate and other number containing the mass transfer coefficient in the boundary layer. This ratio must be less than 1 if the controlling step is the chemical reaction and the fast step is the transport of the protons in the boundary layer. This means that the numerator, containing the reaction rate is lesser than the denominator, with the mass transfer coefficient. This condition is meet with pH values lesser than 4.

The next step is to get the values of the terms involved in the inequality and the following procedure was followed. Particle diameter of the agglomerate of TiO_2, m = cm. To get the value of the mass transfer coefficient, the general mass balance was analyzed for a spherical particle in a fluid, without reaction, no convection. The differential equation was solved to find the concentration profile, using as boundary conditions that in the surface of the particle, the concentration is C_R and for an infinite radius, the concentration is cero. Using Fick law, the mass transfer of the protons was calculated to calculate the molar flux.

The mass transfer coefficient is defined as the proportionality constant between the flux and the difference of concentration between that in the surface and in the fluid. Then, with an equation of mass analogous to the Newton´s cooling law, we deduct the mass transfer coefficient and calculate it with the given information.

Using one concentration profile of protons, the reaction rate was calculated, and substituting the proton concentration at pH = 4 in the rate , the value was 0.389 (less than one). In this way, the mass transport is rapid and the reaction rate is slow and is the controlling step. This occurs for pH less than 4. For pH bigger than 4, this inequality is not meet, and the mass transport controls. Then, at pH lesser than one, the ions transport in the boundary layer is enhanced and increase the reaction rate. This is corroborated by the experimental data.

Problems two.

We are aware the molecular hydrogen is produced on the surface of the catalytic particle, then, it needs to cross the boundary layer surrounding the particle, to transport in the liquid phase, to cross the liquid – gas interphase, then to transport in the gas phase until reaching the syringe needle where the sample is collected. Then, the goal is to calculate the molecular hydrogen that is in the gas phase, using the existing theory.

The catalytic particle is surrounded by a boundary layer which offer resistance to the proton mass transport. Fick law including diffusion and convection and a mass balance in a differential element are used to obtain the differential equation describing the change of molar fraction of the protons. Solving it and using the appropriate boundary conditions, the molar fraction profile is obtained, and then the expression to calculate the molar flux in the limit of the boundary layer. This equation allows calculate the proton flow crossing the limit of the boundary coming from the water, in direction to the catalyst surface. The corresponding data are substituted. Certainly, there is uncertainty in several data assumed to do the calculations.

According to the stoichiometry, it is possible to determine the flux of molecular H_2 in the limit of the boundary layer, and then to predict the transport in the liquid phase to reach the gas – liquid interphase. Raoult law is used to calculate the molar fraction of molecular hydrogen in the liquid, assuming the flux is constant from the limit of the boundary layer to the interphase liquid – gas. It is assumed the liquid is quiet and the convective term is neglected. An equilibrium curve is used to relate the molar fraction of hydrogen in water and in nitrogen.

To analyze the transport in the gas phase of nitrgen, it is assumed the flux is constant and is the same than that in the liquid phase. Besides, it is considered the gas is quiet and there is not convection. With the Fick law, the molar fraction of the hydrogen is calculated in the gas phase, and it is compared with that measured by the gas chromatograph. The error is round the 80%, and this can be attributed to the uncertainty in some data assumed due the lack of information.

In this way, the prediction of the experimental results obtained in this research line, use a procedure and calculations relatively simples than can be used in the programs of mass transport in undergraduate program of chemical engineering.

542645 Marketing is Not Bragging: How to Articulate Your Value to Advance Your Career

Abstract:

AIChE Career Services is sponsoring a workshop to help you improve your job seeking skills.

Marketing is not bragging: How to articulate your value to advance your career

In this workshop, we will discuss how to identify your value, and appropriately self-promote and market this value to clarify what you can do to help potential employers and decision-makers. Learn how to craft and deliver multiple elevator pitches and gain insight into specific strategic marketing channels that you can leverage to ensure your brand (your promise of value) is amplified to the right people.

527056 A Call to Service â€“ Areas of Focus Where You Can Contribute and Make a Difference to Engineering

Abstract:

In this work, I will provide brief insights as to why (I think) I have been involved in service to my profession throughout my career. However, I will mostly focus on a call to the audience for service in a variety of areas (with justification as to why I think these are important areas). These include the following:

1. Approaches to address the lack of educational experience for faculty candidates in engineering and what you can do about it.
2. Safety, now more than ever, in the chemical engineering curriculum and where you can play a part.
3. The next generation of engineering students â€“ opportunities to attract and serve them.

My perspective as Associate Dean for five years and Dean for two years provides a broader context than only chemical engineering, but the challenges and opportunities are similar.

Enough time will be provided for questions and answers so that those in the audience can either delve deeper into a specific area (and my perspective) or provide their own thoughts.

524815 Off-Road Professing: Creating Positive Change in Curriculum and Department Culture from Outside the Tenure Track

Abstract:

The presence and role of non-tenure-track faculty on university campuses are rapidly growing. However, this role is often ill-defined and in flux for many chemical engineering departments. While non-tenure-track positions often leave lecturing faculty with less security and without the full powers of department governance, many opportunities remain to define a role which can enhance entire curriculum and department culture. Such positions, focused on engineering education, may even present opportunities to affect positive department change that would be more challenging for those working to establish tenure. In this talk, three key areas of lecturing faculty impact will be highlighted: 1. Curriculum research and development; 2. Student advising; and 3. Community outreach. Successful methods for effecting change and securing a department role for non-tenure-track faculty in each of these areas will be discussed.

537139 Enthalpy, Entropy, Confused Students & Overworked Faculty-What We'Ve Learned about Students' Misconceptions in Thermal Sciences & Faculty Adoption of Effective Practices

Abstract:

Our intuitive beliefs about how the universe works are not always borne out by science, and these beliefs can be quite resistant to change. A lifetime of experience that concrete floors are "colder" than carpet is not easily overridden by a lecture on thermal conductivity. In this talk, I'll discuss how we worked the challenge of students' misconceptions in thermodynamics and heat transfer. We started by working on measures to assess change in students' conceptual understanding, then created and studied inquiry-based activities to repair students' misconceptions, re-imagined those activities in easier-to-use formats, and studied the adoption of those activities by faculty. Our current focus is on faculty adoption, because the best educational activities in the world are destined for low impact if faculty don't use them. I'll conclude with our heuristics for what makes an educational innovation potentially 'catchy' for faculty.

537452 Act Now! Operators Are Standing By: Student Pitch Videos

Abstract:

Video has become a pervasive communication method in the United States; over half of the US population accessed YouTube at least once last year, often for hours at a time. Employers interviewing our graduating seniors often request video as part of the interview process. It's likely many of your students frequently share short video through Instagram or Snapchat (occasionally even in class). Simple video production using cell phones is now within the capabilities of most student teams. Helping our students to become better producers of accurate, succinct, and engaging video media is increasingly important.

We have developed a short assignment that requires students to summarize their class project as a 1-2 minute "pitch video." We have successfully deployed this assignment in our senior design sequence, thermodynamics, as well as elective courses. Students' video production is scaffolded with examples, guide sheets, and pointers to example media and supported with a limited number of iPads and microphones for loan. This scaffolding and support approach has resulted in better quality videos as well as in enhanced student reflection on their projects.

538190 By Students for Students: Using Course Projects to Create Learning Materials for Future Classes

Abstract:

In order to improve student understanding of course concepts, as well as to increase the range of learning material available for students, a course contribution project was introduced in a Transport II course. This project required students to find potentially fun and effective means to communicate topics related to heat and mass transfer, with the project deliverables providing new learning tools that could be used to help other students learn.

Students were given four potential project types to select from: writing sample exam problems, for which both problem statements and solutions were required; producing a 5-minute video; writing and drawing a 10-page comic; or developing an experimental module that could be replicated by other students. All of the projects were required to focus on concepts related to heat and/or mass transfer that were addressed in the course, including conduction, convection, fins, heat exchangers, and separations equipment. The intended audience for each project output was the studentsâ€™ classmates, essentially meaning any Transport II student.

With the permission of the students who completed the project, the deliverables were shared with students who took Transport II in the following semester, thus providing new problems and explanations to learn from. In addition, students who completed the project claimed to gain more in-depth understanding of the course topics related to their projects.

Assessment of the projects over several semesters, as both a deliverable by students and a learning tool for students, will be further discussed.

517413 Using Peer Review to Improve Student Learning Outcomes Associated with Ill-Structured Problems

Abstract:

There is a gap between the well-defined engineering problems that students solve in their coursework and the ill-structured types of problems that engineers encounter in professional practice. Some obstacles that prevent a more routine use of ill-structured problems in engineering courses is that assessing student solutions and providing specific feedback to learners takes more time for open-ended problems. One way to overcome these barriers is to get the learners involved in the assessment and feedback process. This presentation will focus on the use of peer review as a mechanism to help individual students and student teams improve their problem solving approaches. Specific peer review cases will demonstrate how to structure a review process that drives students toward writing meaningful evaluations for the work of their peers and includes time for students to reflect on their work and how to improve it. Evidence showing the positive impact of peer review on student learning outcomes will be shown.

536688 Increasing Student Interaction with Anonymous Polling

Abstract:

Several of the sessions at Summer School made use of anonymous polling websites to facilitate interaction among participants. Polling is an easy way to verify student understanding of concepts discussed in class. In addition, polling allows for questions to be posed that require individual critical thinking, as students are able to think independently prior to offering their own opinion. In large classes or classes with shyer students, the ability to have all students participate is key to student learning and engagement. Polling students through the use of clickers has been utilized for over a decade, but clickers can be costly. Polling websites enable students to turn their phones or computers into clickers either for free or for a small fee. In this talk, the use and effectiveness of polling using Poll Everywhere in a variety of chemical engineering courses will be demonstrated.

525211 Get in Pairs and Roll the Dice!

Abstract:

One of the most useful ideas I took away from last yearâ€™s ASEE summer school was the way to break up lecture so studentâ€™s brains could reset and be ready to receive new information. During the plenary sessions, I noted a contrast in how I felt when 1) we periodically broke up into groups to discuss topics during the talk, and 2) the speaker just talked on. For situation 1) I felt energized and interested, while for situation 2) I felt drained and bored. For this reason, I decided to incorporate one to three minute breakout discussion questions into my lectures. I then acquired some 10-sided dice and rolled them to see who would have the opportunity to answer the question. Eventually, I decided to make it more participatory by having a student roll the dice. Although I initially had to encourage students to get in pairs and talk, they got used to the regime and I had no complaints about it. In this presentation, I will discuss student comments and changes in learning outcomes.

542620 The Impact of Shale Gas and Oil on the Chemical Industry

Abstract:

Lower alkanes and aromatics are very useful as both fuels and chemical raw materials. Occasionally supply disruptions constrained availability but also encouraged investigation of alternative feedstocks and chemistries. Recent new technologies for the production of unconventional gas and oil sources from tight shale formations significantly increased availability of both methane and condensate, radically altered process economics, resulted in the construction of much new intermediates capacity, and multiplied the ways traditional and alternative chemistries might be exploited. This presentation will outline historically how natural gas components became useful as chemical feedstocks, how government regulations occasionally led to perturbations in availability and price, how supply crises led to the development of alternative chemistries, catalysts, and ultimately the exploitation of shale resources. More specifically, will be discussed how excess coproducts, especially condensate, from increasing shale gas and oil production for energy are having significant unexpected impacts on the chemical process industries and opportunities for new process chemistry research.

541186 Sustainable Energy and Chemicals: Past, Present, and Future

Abstract:

The last 40 years has seen the U.S. energy situation vacillate from gasoline rationing and concerns over Peak Oil, to projected energy and feedstock prices that were noncompetitive in global markets for chemicals production, to the economic revolution of frack gas and light tight oil. The importance of low cost energy and feedstocks in driving economic opportunity will be discussed relative to this history, along with the importance of sustainability in decision making to gain stakeholder buy-in and acceptance. Strategies for managing career paths and competencies despite changing targets and high uncertainty will be addressed.

542618 Disruptions: What the Future May Hold

Abstract:

Disruptive forces are all around us. They change the way we communicate and work. They change the way we develop goods and services. They also change the types of products that are manufactured. Examples such as the way in which we produce and consume energy or the feedstocks we use to produce product have seen many changes over the course of the last few decades, and these disruptions are expected to continue. The chemical engineering discipline and the expectations for a chemical engineerâ€™s skills has also changed and adapted as a result of these disruptive forces. A discussion of the broad themes associated with these forces and the implications that they may have to chemical engineers in the future will be discussed.

542621 Geopolitical Factors Influencing the Evolution of the Chemical Industry

Abstract:

The basic chemicals and plastics industry is built on a simple principle: find a cheap feedstock and convert it into something more valuable. However, the right feedstock depends on place and time. Certain policies and actions of various governments have influenced the evolution and profitability of the chemical industry during the past 50 years, sometimes in surprising ways. For example, high oil prices and US government funding encouraged the development and commercialization of coal gasification technology during the late 1970s and early 1980s. This enabled China to develop a competitive basic chemical industry using coal as a feedstock during the last 25 years. Similarly, an abrupt increase in oil prices during the first decade of the 21st century encouraged the present shale gas revolution, causing a major shift in the economy of producing basic chemicals and polymers in the US. If history is our guide, then what governments choose to fund and regulate may be as important as market forces in determining the shape of our industry and profession in the future.

542622 Agility & Resilience: How to Maintain Career Competitiveness in the Changing Chemical Industry

Abstract:

The world around is changing. The shift to protectionism, global instability and price volatility of raw materials and feedstocks has the potential to wreak havoc on the chemical industry. With the revolution of shale gas in North America and the sudden drop in feedstock costs, the chemical sector in the US has reasserted its historic dominance. Learn how the role of a chemical engineer can contribute to industry growth and resilience, while adapting to allow for personal agility throughout the career life cycle.

540898 Practical Adaptations for Teaching Students with Disabilities

Abstract:

In this talk Ms. Neybert presents personal experiences as a student, teacher, and curriculum designer with multiple disabilities. Economically friendly classroom and laboratory accommodations for students with disabilities will be discussed, with live demonstrations of selected accessibility accommodations. Finally, learn how investing time into a student with disabilities could produce a star colleague for the future.

531424 An Interactive Tool for Learning Spreadsheets

Abstract:

Spreadsheet training can be found at many, many websites for at little or no cost, however, almost all tools are demonstrative, i.e., showing the student how to use a certain version of a specific spreadsheet program. While spreadsheets are very important to many jobs that a chemical engineer performs, spreadsheets have become a commodity, which is analogous to filling up at any brandâ€™s gas stations to keep a gas running. Whether using Microsoft Excel, Google Sheets, or Apple Numbers on a computer, tablet, or mobile device, the basic functions of spreadsheets holds for all. Using an interactive textbook from zyBooks provides a technology to actively learn how to create and use a spreadsheet. Common formulas, functions, and features of spreadsheets can be presented, interacted with inside the interactive textbook while translating to the learnerâ€™s preferred spreadsheet program. Animations and question sets streamline blocks of text or static screenshots used in other textbooks or websites. Learner interactions are recorded, so instructors can award participation and homework grades. Concepts covered include: formulas, functions, charts, trendlines, solver, error, statistics, interpolation, and more. Over 100 auto-graded homework problems with randomized numbers and content are being used by students during Spring 2018. Results from tens of thousands of student interactions from a single cohort of student preliminary show a high reading rate (90%) and success on auto-graded homework problems (87%). This work will build upon work presented at the 2018 ASEE Annual Meeting.

532783 The Emulsion Lab: An Industry Relevant Experiment for Senior Unit Operations

Abstract:

The senior unit operations course in the chemical engineering curriculum presents an opportunity for students to gain a hands-on experience with small-scale equipment commonly used in industry. Typically, unit operations consist of traditional pieces of equipment such as distillation columns, flow through pipes and fittings, and heat exchangers. Although these unit operations provide opportunities for students to connect with material taught in the classroom, there are still a large number of concepts that students are not prepared for upon entry to industry such as emulsion technology. Emulsions are found in a wide range of industries and would be beneficial to incorporate into the classroom. The envisioned open-ended emulsion experiment consists of vegetable oil, water, lecithin, Tween 80 and sodium dodecyl sulfate and salts to investigate the effect of surfactant concentration, type of surfactant and ionic strength on emulsion stability.

534067 Engaging Engineering Students Using Memes

Abstract:

Whether class lectures are 1 or 4 hours long, it can be difficult for students to stay focused on the materials at hand. One way to maintain student attention is to include short cartoons or memes in lecture slides. Memes are usually represented by humorous images, videos, or text, and they are often copied and spread â€˜viralllyâ€™ through the Internet. As memes exist for nearly every concept, they can be used to showcase chemical engineering in an entertaining manner, and there are many benefits to including memes in course discussions. The use of humor in memes is meant to grab peopleâ€™s attention, therefore students are more likely to remember course materials if they can visualize these concepts with pictures or funny ideas. Memes are also very popular with college-age millennials, therefore they are a great way to relate to students and to engage them to think about engineering concepts outside of class. In this talk, the use and effectiveness of memes in a variety of chemical engineering courses will be demonstrated.

532494 Integrating Sustainability Principles into Chemical Engineering Core Courses: A PBL Approach

Abstract:

Today, there is broad consensus about the urgent need to integrate sustainability principles into chemical engineering curricula. However, attempts to do so raise important questions: (1) how to accommodate these principles in a crowded syllabus or curriculum? and (2) how to design separate courses to teach sustainability with limited resources?

The current work proposes the implementation of a project-based learning (PBL) approach to integrate sustainability principles into chemical engineering core courses at The Polytechnic University of Puerto Rico (PUPR). Currently, this approach is being piloted in two courses: chemical engineering thermodynamics and mass transfer operations, which are taken by the students during their junior year.

In the thermodynamics class project, each team is tasked with the design of an absorption refrigeration system to recover wasted heat from a specific process. To achieve the project objectives, the students are asked to use energy and mass balances, in addition to some thermodynamics concepts such as coefficient of performance (COP), isentropic efficiency, and the first and second laws.

For the mass transfer operations class project, each team is asked to design three different absorbers (single unit, multi-stage and packed column) to recover SO_2 from an industrial flue gas to subsequently convert it into a value-added product. As part of the project, the students apply core concepts such as the operation and pseudo-equilibrium lines, and the fractional approach to flooding.

In both projects, the students must consider the following aspects: (1) sustainability principles, (2) the existing environmental regulations and (3) the social and economic impact of the proposed design.

Typical summative assessment methods are used to evaluate the impacts of these PBL activities on the student learning outcomes. In addition, the studentâ€™s perception of learning is assessed with the Student Assessment of Learning Gains (SALG). Also, student self-efficacy is evaluated using the 8-item self-efficacy subscale from the Motivated Strategies for Learning Questionnaire (MSLQ).

This pilot study provides valuable information to the academic community, in terms of how to design and develop activities to expose chemical engineering students to sustainability issues while learning specific principles of the field.

533467 Revolutionizing Engineering Education at Oregon State University

Abstract:

This poster presents our efforts towards *revolutionizing* the undergraduate learning environment in the School of Chemical, Biological, and Environmental Engineering at Oregon State University (CBEE). While industry has long called for changes to how engineers and computer scientists are prepared, much of the core coursework remains virtually the same as it was 50 years ago. To address this mismatch between industry and education, the National Science Foundation has called for proposals that aim to *REvolutionize Engineering and Computer Science Departments* (RED), serving as leaders in institutional change. We describe early results and ongoing work in a RED project in the School of Chemical, Biological, and Environmental Engineering (CBEE) at Oregon State University. We seek to re-situate the faculty/staff/student community and curricular/co-curricular student learning experience. Our work is centered on the creation of a holistic, inclusive, and professionally-based learning environment for our students. We are working to promote change through establishing a culture of inclusion and shifting student learning environments from highly sequestered activities to more realistic and consequential work that more closely resembles actual engineering workplace. These efforts require an important and concurrent shift in unit culture (e.g. values, norms, policies and procedures).

537946 Collaborative Writing "Wiki Tool" in a Chemical Engineering Laboratory

Abstract:

This work will introduce a collaborative writing pedagogical tool for use in an upper-division chemical engineering laboratory course. The chemical engineering curriculum ubiquitously includes a capstone course (in either the junior or senior years) characterizing industrially relevant instrumentation, often referred to as unit operations (UOs). This work proposes the use of a Wiki tool to replace student reports, initially for a subset of the UOs. Instead of submitting reports to an instructor, as is typical, students upload their work as a page on the course management system (CMS, e.g., Canvas, Blackboard, etc.), visible to all enrolled in UOs. Subsequent teams studying the same UO provide peer feedback on the original report, validate the data collected, and perform additional experiments investigating the effect of parameters not yet studied; finally, for their final report, they will edit and amend the original document on the CMS with sections related to their study of the UO. It is hypothesized UOs using this pedagogical tool will observe that: (i) the quality of uploaded reports increase, given their responsibility to the next team, (ii) the quality of peer feedback increases, (iii) students gain a more thorough understanding of the UO, and (iv) students understand how to verify others' results. In this poster, methods and preliminary data will be discussed regarding this work currently in progress.

514627 Developing Standards for an Operations Center Process Safety Educational Exercise Using Simulators

Abstract:

Columbia University Chemical Engineering and Simulation Solutions Inc four-day Chemical Engineering Operations Center educational exercise combines computer simulation with a role player exercise and classroom instruction in order to educate students on chemical engineering process safety and to provide a unique chemical engineering process safety experience. In 2018, this academic and industry team addressed the need to further develop the educational exercise by an added component of standards for an Operations Center Simulated Chemical Engineering Process Safety Education Exercise. With multiple years of integrating simulation of chemical engineering processes, the CHEN E4501 Chemical Engineering Process Safety Course uses case studies, computer aided design, and simulation in the course content covering instructional objectives. Course content is also on the education and training of operators using simulation. The results of the 2018 lesson sequence on operator training coupled with an operations center role player exercise sponsored by Simulation Solutions Inc. and the methods used to evaluate student performance as related to safety outcomes are reported on here. This presentation will focus on development of standards for evalution of the operations center exercise combined with simulation.

539227 Development of "Smart Materials" Master's Degree Program Module for Chemical Engineers

Abstract:

Global trends in the development of science and industry, include the concept of "smart materials" as an integral part of the global knowledge-based industry in the horizon of the next 15-20 years.
Thisproject is aimed at developing a group of training courses "Smart Materials" for the Master's program "Physicochemical Foundations of Innovative Technologies for Supermolecular Systems", which is currently taught Kazan National Research Technological University, Russia.
The main idea of this educational initiative - courses, developed in English, will become a catalyst for the internationalization of the Master's program and its further development as a complete set of components focused on international students.
The target audience of the training course is represented by Russian and foreign students of engineering specialties, specializing in the field of chemical engineering, materials science, nanotechnology, biotechnology, and etc.
The educational product developed during the project implementation will also be used as an independent training module for the training of foreign students, for the implementation of short-term mobility programs such as summer schools, and for implementing network educational programs with universities in Europe and South-East Asia.

The volume of the smart materials market will reach 70 billion US dollars by 2020 (according to Zion Market Research).
Accordingly, the next 15-20 years, we should expect a steady increase in demand for engineering professionals who have competencies in the field of nanotechnology, materials science, smart materials and their practical application in promising technology industries.
Within the framework of this project, three training courses in the field of smart materials "smart materials" were developed.
The titles of the courses:
1. Introduction to smart materials: biosystems and nanosystems
2. Technology of smart materials
3. Research methods for smart materials
Each course is a semester course and offers three credits.

For the course "Introduction to Smart Materials: biosystems and nanosystems," a lecture block was developed in English. The unit will include video and audio components read by the native speaker from among the international partners of Kazan National Research Technological University.
The course "Technology of Smart Materials" is also represented by a lecture block in English. It is followed by a practical unit - a set of five laboratory experimental practicums. Laboratory practicums are based on five advanced experimental techniques: photon correlation spectroscopy (dynamic light scattering), contact angle measurement, spectrofluorimetry, IR spectroscopy and spin-coaching.

The course "Research Methods for Smart Materials" is to be implemented in partnership with the nanocenter of the Republic of Tatarstan, Russia, providing the access to the most advanced equipment for studying functional nanomaterials.
The lecture materials cover the the following topics:
1. Analytical spectroscopy;
2. Biochemistry and biomedicine;
3. Introduction to Materials Science;
4. Introduction to modeling;
5. Introduction to nanotechnology;
6. Introduction to nanotechnology and innovative materials;
7. Information technology;
8. Research methods for studying nanosystems;
9. Materials for molecular electronics and biomedicine;
10. Materials and composites;
11. Advanced nanotechnology;
12. Project planning and literature reviews;
13. Microfluidics;
14. Production of micro and nanomaterials, "pure" technological processes;
15. Spectroscopic analysis;
16. Functional materials;
17. Characterization of materials;
18. Chemistry of materials;

Graduates of the master's program will obtain the following academic, scientific and intercultural competences:
- readiness for intensive professional communication;
- readiness for project work in international teams;
- expertise in developing new materials and technologies for innovative materials;
- proper selection of modern materials and technologies in the field of smart materials;
- ability to search, process and systematize scientific and technical information in the field of nanotechnology;
- the ability to independently learn new methods of research, to change the scientific and scientific and production profile of their professional activities;
- the ability to predict the properties and behavior of polymers and nanosystems in the processes of self-organization at the molecular and supramolecular levels;
- the ability to comprehensively analyze the existing and prospective market of smart materials.

Development and implementation of his module will be followed bycreating a new Master's degree program.
A new (for Russia) methodological element proposed within the framework of this project is implementation of capstone projects. Project teams will incklude students as well as Russian and foreign (remote consulting) scientists and industry representatives to work on practical projects in the field of innovative materials science, which are of practical interest to chemical engineering companies.

The evaluation of the project's success will be based on the following criteria:

- the number of Russian students enrolled to this Master's program after the introduction of the courses "Smart Materials";
- the number of internationsl students enrolled to this Master's program after the introduction of the courses "Smart Materials";
- the number of new teaching materials developed for the implementation of this module ;
- the amount of funds raised from the Master's Program;
- the intensity of academic mobility projects with international partners;
- English language proficiency demonstrated by Russian students attending "Smart Materials" program courses;
- attention to program graduates demonstrated by Russian and international employers.

The project was supported by the grant provided by Potanin Foundation for the development of Master's degree programs.

529821 The Link between Spatial Visualization and Chemical Engineering Problem-Solving

Abstract:

The problems of the future are complicated and interconnected. For example, how do we ensure an increasing world population has access to clean, plentiful water? There are many competing needs for the water itself, from agriculture to sanitation to personal consumption, but in addition, there are impacts on the environment to be considered as well as societal impacts between the countries in the developed world and those in the developing world. Not to mention the merely technical issues, e.g., how do we draw the saline out of seawater? This is the messy realm occupied by engineers. They are expected find a path through the complexity and interconnectedness of this and the many other grand challenges of the future. They are expected to be accomplished, insightful problem solvers.

In addition to their reputation as problem-solvers, engineers are also known for their excellent spatial skills. In tests with more than 30,000 professionals, engineers demonstrated the highest level of spatial visualization skills, followed closely by architects and other STEM professionals. A recent study showed a strong correlation between spatial visualization skills and creativity and technical innovation.

It is posited that possessing advanced spatial abilities allows an individual to construct robust mental representations of problems. These representations are known to be critical in solving all manners of problems but particularly in the case of "insight" problems where overcoming an impasse is necessary. Dealing with an impasse in a problem often necessitates re-representing the problem situation so that a different approach may be adopted.

Approximately 60 undergraduate students in chemical engineering were administered a test of spatial cognition (the Mental Cutting Test, or MCT). They were also given 12 â€œtypicalâ€ chemical engineering problems to solve. Problem sets were administered during a class period in Thermodynamics, a third-year course in the program, but the topics tested in the problems are typically covered in a second-year course in chemical engineering. Since the students were all given the same amount of time, those who were better at problem-solving would likely solve a larger number of problems compared to weaker students.

A strong positive correlation ($R=0.59$, $p<0.00001$) between spatial skills test scores and the number of problems successfully solved by the students was found. Problems where spatial skills appear to play a role were identified and will be further described in the presentation.

Previous studies have shown links between spatial skills and overall success in engineering. Through this work, specific types of problems in chemical engineering were identified where spatial skills appear to play a role in solving them. This has implications in curriculum design within undergraduate chemical engineering programs.

536404 Photobioreactor Design andÂ Biodiesel Synthesis

Abstract:

The negative environmental impacts of combustion are well-known. As part of our introductory-level chemical engineering design and innovation lab, students learn about biodiesel as an alternative, cleaner, carbon-neutral fuel for diesel engines. Students are taught about the possibilities and difficulties associated with growing algae on a large scale, extracting the oil from it, and converting it to biodiesel. In the lab, students work in teams of three and are tasked with designing and constructing a photobioreactor for optimizing the growth of algae while minimizing water evaporation. Students track the algae growth over the course of four weeks by measuring its optical density with a spectrophotometer each day and obtain a cell count using a hemocytometer at the beginning and end of the project. While the algae are growing, the students synthesize biodiesel from their choice of a wide variety of other starting oils, such as vegetable, corn, olive, canola, peanut, avocado, and more. Students then analyze their biodiesel and traditional diesel for flame temperature, burn rate, qualitative viscosity and density. They further analyze their biodiesel using a variety of analytical equipment, including an FTIR-ATR, UV-Vis spectrophotometer, and refractometer.

Each group of students submit all of their collected data from the two projects and are then tasked with analyzing the data from the entire course, combining their own groupâ€™s data with that of their peers. Students must first assess the reliability of the various data sources and determine, with justification, which data they believe is inaccurate and should be omitted from the analysis. Their objectives are then to identify which photobioreactor designs are most effective and which starting oil yields a biodiesel product most similar to traditional diesel.

529530 Impact of Online Numerical Response Questions on Student Learning in Parallel Sections

Abstract:

Thermodynamics is taught at The University of Akron as a 4 credit course in the Spring of the Sophomore year. The course follows on the Material and Energy Balances course based on the text of Felder et al. In recent years, growing class sizes have led to offering the course in two sections taught in parallel with different instructors. For the past two years (2017 and 2018), the course was taught with the same textbook (Elliott and Lira, 2012), and identical tests and final exam on identical dates. The course implements three projects, a weekly workbook that emphasizes concept tests and practice test questions, standard weekly homework problems, and an average to two 10 minute video assignments per lecture. In 2017, both sections applied a weekly in-class 20 minute quiz, but "Section E" also included in-class 15 minute quizzes designed to check whether the students had watched the videos. The result in 2017 was that both sections performed practically the same on all three tests and the final exam. In 2018, the in-class "video check" quizzes in Section E were put on-line in the format of numerical response questions with randomly generated prompts and formulas generated to automatically grade student responses in real time. The on-line "video check" quizzes offered several benefits at the outset: saving class time, no time pressure, multiple attempts (usually 2). More importantly, significant differences were observed between Section E and "Section Z" on test performance. Section E performed 12% better than Section Z on Test 1, and 15% better on Test 2. This corresponds to the difference between a 72% average for one section and a 57% average for the other, as an example. For Test 3, both sections were assigned the same video check quizzes, but no results are available for Test 3 as of this writing. Student perceptions were also surveyed through the same quizzes and these will also be presented and interpreted. After many years of applying many technologies and new methods in the classroom, no technique has had such a significant and demonstrable impact in such a short time. Creation of the question database is somewhat labor intensive, but it is available as a D2L package that is also portable to Blackboard. Porting has been tested across D2L platforms only.

530350 Use of Numerical Software in Education and Research

Abstract:

Until the early 1980's, main frame computer use in Engineering Education and Research involved mainly FORTRAN programming and less frequently CSMP programming coding. With the introduction of personal computers in the early 1990's, several types of interactive programs for numerical solution of engineering problems developed. The intention was to replace the old time consuming and often inaccurate solution techniques (such as 1. Analytical solutions that may include model simplification by neglecting less important terms and model manipulation to bring it into a solvable form. 2. Short-cut solution techniques where the problem is replaced by a simpler one that can be solved, 3. Trial and error solution techniques and 4. Graphical solutions) with numerical solution techniques. Over the years, several types of numerical solution packages have been developed and utilized: 1. Multi-purpose general mathematical packages such as POLYMATH[1], MAPLE[2], MATLAB[3], and Mathematica[4], 2. Spreadsheet programs, such as Excel[5], 3. Process Simulators, such as Aspen and HYSIS, 4. Computational Fluid Dynamics Packages and 5. Molecular Modeling.

About 15 years after the introduction of numerical software packages, surveys regarding the extent of their use in engineering education were conducted. One such survey was carried out by Jones (1998). His conclusion was that: "Across the country, computers are usually not used effectively in undergraduate engineering science courses. Often they are not used at all. Problem solving approaches and calculation methods are little influenced by the availability of computers." Dahm et al (2002) reached the following conclusions regarding the use of process simulators at that time: "Chemical process simulation is currently underused in the chemical engineering curriculum at many schools . . . many respondents acknowledge that the role of simulators could be beneficially expanded in their curriculum."

Our objective in this paper is to investigate the penetration of numerical software into engineering education and research in the last 20 years and summarize the current status. Our investigation at this time is limited to the utilization of the POLYMATH package.

The roles of the various numerical software packages in CHEG education, research and practice have been discussed, for example, by Shacham et al. 2009 and Shacham and Cutlip 2004. In this presentation, the extent of the use of the POLYMATH package (that was developed by the authors of this paper) over the last 20 years is reviewed. The educational part of the review is based mainly on textbooks where POLYMATH is utilized extensively for problem solving (Fogler, 2006, Cutlip and Shacham, 2008, Hagen, 2015 and Razdolsky, 2017). The "research" related part of the review utilizes the "Google Scholar" search for "POLYMATH Software" in the scientific literature.

The first version of POLYMATH was developed for the PLATO educational computer system (Shacham and Cutlip, 1982). The first commercially available version for the IBM PC was published in 1993. Versions of this program have been continuously updated and are now available now for the current Microsoft Windows and the Android operating systems.

The most widely used capabilities within the POLYMATH package include: 1. Single and several simultaneous nonlinear algebraic equations (NLE's) solver, 2. Simultaneous ordinary differential equations (ODE's) solver, and 3. Linear, polynomial multiple linear and general nonlinear regression.

POLYMATH was initially intended for educational use, and as such it has several user friendly features. These features include 1. Menu based (instead of command based) program control. 2. Notation and format of the equation entry are very similar to the notation and format of the mathematical model. 3. The program reorders the equations according to the computational sequence and detects implicit relationships among variables. 4. Several debugging aids identify possible programming errors. This last user friendly feature requires keeping the problems to be solved below a certain complexity level (Single-Model Single-Algorithm problems) that are adequate for undergraduate education. Further discussion of more complex problem types (such as Multiple-Model Multiple–Algorithm) and the way of POLYMATH utilization to help solve such problem is discussed for example by Cutlip et al. (2009).

The first textbooks that included POLYMATH numerical solutions to most of the examples presented started to appear toward the end of the nineteen nineties starting with Fogler's "Elements of Chemical Reaction Engineering" (first in 1998 in the 3rd edition and now in the 2016 in the 5th Edition) and Cutlip and Shacham's book "Problem Solving in Chemical and Biochemical Engineering with POLYMATH, Excel, and MATLAB" (first in 1999 and now in 2007 in the 2nd Edition). Articles mentioning the use of POLYMATH in research (non-educational) oriented journals started to appear in 2000.

In the paper by Lueking et. al (2000), the authors (from the department of Civil and Environmental engineering) used the nonlinear regression program of Polymath to determine the optimal parameter values of a Michaelis-Menten equation using experimental rate data. In the paper by Satrio et al. (2000), the authors (from the ChE department) used POLYMATH to determine reaction rate data from measured slopes of octyl acetate concentration vs. time plots.

Beginning with year 2000, the number of publications in which use of POLYMATH is reported have kept increasing continuously reaching about 50 publications per year in 2017. In 2018 there were (as of the end of March) 17 such publications. Details of these publications are provided in Table A-1 which is available in the full version of this Abstract which can be found at:
ftp://ftp.bgu.ac.il/shacham/AIChEMeet18/Abstract18.docx .

Among the publications listed in Table A-1, there are educational and research articles, conference papers and one US Patent application. The authors are from 13 different countries, and they also belong to different academic departments: Chemical, Biological, Pharmaceutical, Environmental, Petroleum, Electrical and Computer Engineering, Chemistry and Agriculture. There is also a wide range of subjects investigated. In educational application, POLYMATH is most widely used for solving systems of ODEs obtained from material and energy balance on chemical reactors or similar equipment. In research, the nonlinear regression tool of POLYMATH is used most often, typically for determining optimal parameter values for rate expressions.

This historical study demonstrates that the use of numerical software for engineering and scientific problem solving is continuing to increase. Solving numerical problems should be finding many applications in STEM areas of study.

In the presentation, some examples will be shown in more detail. The selection of the numerical software package most appropriate for a particular problem, based on the complexity of the problem, will also be discussed.

References

1. Cutlip, M. B. and M. Shacham, "Problem Solving In Chemical Engineering with Numerical Methods." Prentice-Hall, Upper Saddle River, New-Jersey, 1999
2. Cutlip, M. B. and M. Shacham, "Problem Solving In Chemical and Biochemical Engineering with Polymath, Excel and MATLAB." Prentice-Hall, Upper Saddle River, New-Jersey, 2008
3. Cutlip, M. B., N. Brauner. and M. Shacham, "Biokinetic Modeling of Imperfect Mixing in a Chemostat –

an Example of Multiscale Modeling", Chemical Engineering Education, 43, 243-248, 2009
4. Dahm, K. D. Hesketh, R. P. and M. J. Savelski, "Is Process Simulation Used Effectively In Che Courses? ", Chemical Engineering Education, 2002
5. Fogler, H.S., Elements of Chemical Reaction Engineering, 3rd Edition, Prentice-Hall, New York, 1998
6. Hagen, J., "Industrial Catalysis: A Practical Approach," 3rd edition, Wiley-VCH; 2015
7. Jones, J. B. "The Non-Use of Computers in Undergraduate Engineering Science Courses", J. Engr. Ed., 87(1), 11, (1998).
8. Kyle, B. G., "Chemical and Process Thermodynamics," 3rd Edition, Prentice Hall, 1999
9. Lueking, A. D., Huang, W., Soderstrom-Schwarz, S., Kim, M. and W. J. Weber, Jr., "Relationship of soil organic matter characteristics to organic contaminant sequestration and bioavailability", Journal of Environmental Quality; Madison, 29(1), 317(2000).
10. Razdolsky, L., Probability Based High Temperature Engineering: Creep and Structural Fire Resistance, Springer; 2017
11. Shacham, M. and M. B. Cutlip, "A Simulation Package for the PLATO System," Computers Chem. Engng., 6(3), 209-218 (1982).
12. Shacham, M. and M. B. Cutlip, "Enhancing Computer-Based Problem Solving Skills by Combination of Software Packages", Proceedings of the 2004 American Society for Engineering Education Annual Conference & Exposition, Salt Lake City, Utah, June 20-23, 2004
13. Shacham, M., M. B. Cutlip and N. Brauner, "From Numerical Problem Solving to Model Based Experimentation – Incorporating Computer Based Tools of Various Scales into the ChE Curriculum", Chemical Engineering Education, Vol. 43. No. 4, 315- 321 (2009)
14. Tharakan, J., "Developing Creative and Critical Thinking Skills Through Open Ended Design Projects at the Freshman and Senior Level", Journal of Engineering Education Transformations , 31(3), 2018, ISSN 2349-2473, EISSN 2394-1707

[1] POLYMATH is copyrighted by M. Shacham, M. B. Cutlip and M. Elly (http://www.polymath-software.com)

[2] Maple is trademark of Waterloo Maple, Inc. (http:// www.maplesoft.com)

[3] MATLAB is trademark of Math Works, Inc.(http://www.mathworks.com)

[4] Mathematica is trademark of Wolfram Research, Inc. (http://www.wolfram.com)

[5] Excel is trademark of Microsoft Corporation (http://www.micro- soft.com)

530713 A Fully Online Matlab Course for Freshman Chemical Engineers

Abstract:

All chemical engineering students at UC San Diego are required to complete a four credit introductory MATLAB coding course which covers basic programming techniques and engineering computations. Previously the course has only been offered in a traditional lecture format. A fully online, self-paced, and for-credit MATLAB course was designed in collaboration with MathWorks and piloted in Winter 2018. An overview of the course structure, outcomes, and scalability will be addressed, as well as student perceptions and the MathWorks-academia collaboration.

530792 Social and Tactile Augmented Reality in an Undergraduate Chemical Engineering Laboratory

Abstract:

Augmented reality (AR) has the potential to be a versatile tool for education enrichment. We have constructed an AR table to supplement traditional chemical engineering laboratory education. This AR table uses computer vision software to simulate a chemical reactor plant, building student intuition in a hands-on way that cannot be replicated with a "real-life" lab experience.

538126 Using Wiki Technology to Streamline Your ABET Portfolio

Abstract:

Alignment of evidence, artifacts, measurements, assessments, and evaluations against ABET criteria can be cumbersome for faculty to ensure the reviewer sees what they need to see. Specifically, our department utilized learning outcomes different from the ABET a-k outcomes. As such, a translation and alignment of our learning outcomes to the a-k outcomes was a challenge. As such, the portfolio was transferred to wiki technology embedded with hyperlinks and structured more like a digital file structure rather than a series of notebooks. The authors will present how the wiki technology works, how the portfolio was transferred, and how the hyperlinks helped streamline analysis by the reviewer. If time allows we will also share how digital tools like the CATME and the concept warehouse helped provide evidence for some of the learning outcomes.

532204 A Web-Based Database-Driven Assessment Management Tool

Abstract:

Engineering departments routinely perform and document certain tasks that foster a cycle of continuous improvement. Such a process is designed to focus the educational program and maintain its quality. A well designed process requires collecting, organizing, and analyzing information from many people. Cooperation from the faculty is essential to an effective process, and this is assured only to the extent that the process is made simple and efficient. A thoughtful application of modern information technology can foster this outcome. Many departments have developed *ad hoc* solutions through application of off-the-shelf computer-based tools, such as spreadsheets, that really are not that well suited for the task. We describe how we have instead developed a tool from scratch that is tailor-made to address the data-management requirements of the assessment process.

The system operates through a web interface, which enables distributed access for uploading of data and management of the process. Behind the interface is a tightly integrated database, which stores many pieces of information relating to the program, its goals, its performance in meeting those goals, and documentation thereof. All of this information can be contributed and viewed by faculty, and managed by faculty and staff who oversee the process. The collected data can subsequently be analyzed using integrated tools, which facilitate the process of identifying and addressing shortcomings in the program, or conversely, identifying elements that are effective. This system has been deployed to almost all departments in the School of Engineering and Applied Sciences at the University at Buffalo, and it is currently being evaluated by several other chemical engineering departments at other universities.

In this presentation we describe the software system and the assessment framework within which it operates.

536680 3D Printed Centrifugal Pump Impellers: A Unit Operations Experiment

Abstract:

3D printing can be used to create objects simply and cheaply that otherwise might require complex and expensive methods. Such attributes have enabled the development of rapid prototyping applications, custom manufacturing, artistic creations, and myriad other uses.

One potential use of 3D printing in the Chemical Engineering Unit Operations is reported here. An experiment has been developed that requires students to design, print, and test a centrifugal pump impeller. The efficiency of the impeller is evaluated through generation of a pump curve.

The experiment offers several unique learning aspects in that students are required to: 1) assemble a centrifugal pump â€“ albeit a small one, 2) develop proficiency with CAD software â€“ a skill desired by employers, 3) translate fluid mechanics theory into a real design, and 4) 3D print a functioning object.

The experiment has been used for the past two semesters in the lab course. Challenges with developing 3D printing experiments will be reviewed and samples of student work will be shared. Reflections on how the experiment impacts student learning will be offered.

536289 Chemical Engineering Lab for Seniors at United States Military Academy

Abstract:

The United States Military Academy has been ABET accredited since 2012. One of our capstone courses for senior chemical engineering students is CH459, Chemical Engineering Lab. This course has evolved since its inception to now include a total of six laboratory exercises, consisting of 6 lab periods each with an expected deliverable of a poster, executive summary, or full laboratory report. The rubric for each lab includes an independently developed lab experimental design/procedure with a minimum of three successful runs with usable data; chemical engineering process controls Laplace block diagram; data analysis and calculations to include mass and energy balance; Kline-McKlintock error propagation and error analysis; and a full CHEMCAD model of the process with plausible explanations of discrepancies. In addition there is a requirement for a face to face in progress review on the deliverable within 24 hours of completion of the last lab period, which allows for adequate coaching and mentoring of the scientific writing process. During the semester the cadets must also do the SACHE training and during the Distillation Lab they are required to practice the startup procedure using a proprietary chemical engineering controls software.

The way in which the course is managed has allowed our chemical engineering seniors to develop acumen with respect to scientific writing and we have seen anecdotal improvement with respect to the quality of the poster, executive summary, and full lab report deliverables. Lastly, the students enrolled conduct peer evaluations on their lab team members to further their development as leaders and lab team members, and reinforce the concept of working as a member of a team.

528611 Using Water to Engage Community College Students and Increase Graduation Rates

Abstract:

Community colleges are home to diverse populations of students. These students are often non-traditional students that have returned to school after many different experiences. Many of the students do not think of a community college degree as a gateway to engineering or other professional careers (Martin et al (2014) Community College Reviews). They view the community college as a technical training school that provides employment opportunities with a two-year degree. However, many of these students are capable and intrigued, but often intimidated, with the idea of receiving a four-year degree. In order to better serve the students that desire a four-year degree and to increase their knowledge of job opportunities with different four-year degrees, a research project was created between Michigan Tech and Wayne County Community College District. The community college students take weekly water samples over the course of a summer project and test the water for biological and chemical components. This is the first time these students have been exposed to an open-ended laboratory without known results. The students complete the 8-week session by creating a poster that demonstrates research they have conducted on the history of their riverâ€™s contamination and the experimental results they found. The outcome of this project is to determine if open-ended laboratory experiences increase the studentsâ€™ likelihood of obtaining a four-year degree in a STEM field. This assessment is ongoing with five years of data comparing student educational outcomes between those that participated in the summer project to students that completed a biology or chemistry course at the same community college. The students who participated in the project were twice as likely to obtain a 4-year degree, obtain a 2-year degree or be enrolled in a 4-year program. This in-progress work demonstrates engagement of a diverse group of students in the Detroit metro area, including career preparation in science or engineering and a method to increase graduation rates for underrepresented minorities in STEM fields.

541565 25 by 25: Chemical Engineering in the Next 25 Years

Abstract:

In 2008, the centennial year for AIChE, we asked 25 chemical engineers for their visions of the profession from around the US and the world: Nine faculty, eight from industry, and eight post-docs and graduate students. We posed four questions related to each individual's industrial sector or area of research, asking for extrapolation, new impacts on existing sectors, new sectors, and an optional, more open-ended comment on the future of the profession as a whole.

In this 110th year of AIChE, we have revisited these questions with many of the same people. Ten years has seen many developments in our profession, in society, and in our thought-leaders' experiences. Here we report on their perspectives, comparing them to the 2008 visions.

541542 The Future of Chemical Engineering Itself

Abstract:

Chemical engineeringâ€™s future is bright because it is vital to so much of society and industry -- and its future is complicated by its rapidly expanding universe of activities. It must represent depth in core skills centered on the chemical sciences including biology and materials. At the same time, it must be transdisciplinary; it must exemplify the need and value of reaching across disciplines in borderless ways.

One way is to create and disseminate a compelling vision and strategy for chemical engineering, research, and education for the future. AIChE and the National Academies are working together to that end. In 1988, the National Academies published "Frontiers in Chemical Engineering: Research Needs and Opportunities.â€ Its committee of authors outlined a wide-ranging roadmap to turn promising research opportunities into reality, while calling for researchers to embrace new frontiers.

Since 1988, that has happened -- along with many other transformative changes like the Web, fracking, personalized medicine, and social media. A new nationwide effort, aided by international contributions, will address both the identity of chemical engineering and its future directions. Its goal is to ensure that the people in chemical engineering are primed to meet society's needs and rapid changes.

544708 Accelerating Innovation through Academic-Industrial Partnerships

Abstract:

In 2011, The Dow Chemical Company launched the University Partnership Initiative (UPI) to sponsor collaborative research at leading U.S. Universities. This 10-year commitment aims to address fundamental questions in industrial chemistry and chemical engineering with alignment to Dowâ€™s next-generation process technology and product development efforts. To date, this portfolio of projects has generated an impressive array of intellectual property, publications, and numerous examples of the collaborative developments implemented at Dow. This presentation will introduce UPI program and highlight several contributions from our collaborators and leaders in the chemical engineering profession.

540559 Maximizing Uptime, Efficiency, and Safety of Industrial Operations through Early Risk Detection

Abstract:

Achieving zero incidents while maximizing process uptime and efficiency is a key vision for operating companies. An important building block towards this vision is establishment of a proactive risk mitigation culture, supported by effective systems and a strong management. Studies have shown that most of the incidents and unexpected process failures can be avoided if the operating teams get timely information about developing risks and take preventive actions early on. While advances have been made in process monitoring and automation in the past decades, there still remains significant technological and behavioral gaps that prevent plant operations from proactively mitigating process risks.

A modern industrial plant monitors thousands of parameters, generating upwards of 50-100 million data points every day. Thanks to recent breakthroughs in machine learning and artificial intelligence approaches, today there are autonomous systems that can sift through this data and point out meaningful and timely insight. This can help operating teams ascertain process issues that are hidden in the data, long before process variables reach critical levels.

In this presentation, a new autonomous system, Dynamic Risk Analyzer™ (DRA), will be introduced that points out early indicators of process issues using proprietary machine learning. The risk indicators enable operations team to respond and make required process changes, days and sometimes weeks ahead of any alarm, avoiding late stage (expensive) fixes. To realize these objectives fully, a proactive management workflow must also be in place. The presentation will discuss new initiatives and workflows developed by plant operations to implement a proactive culture. Real-life case studies from plants using DRA technology will be presented on how this culture change resulted in increased uptime, efficiency and safety of plant operations.

531309 Gaussian Processes for Hybridizing Analytical & Data-Driven Decision-Making

Abstract:

Surrogate models are widely appreciated in process systems engineering [1]. The typical setting focuses on expensive-to-evaluate, possibly uncertain functions. Examples include: modular process simulators [2], integrated gasification combined cycle processes [3], a carbon capture absorber [4], and many other applications, e.g. [5-8]. Resources are typically limited, so effective decision making requires data-efficient learning.

The data science and statistical machine learning communities typically focus on models learned solely from observed data. But chemical engineering applications may also require explicit, parametric models, e.g. modeling known process constraints, operations constraints, and cost objectives [9]. So, work has integrated semi-algebraic functions with those learned from data [10] or developed semi-physical modeling techniques [11, 12].

This presentation surveys the state-of-the-art in hybridizing analytical and data-driven decision making [13]. We consider three probabilistic modeling applications to these hybrid situations:

Design of experiments for model discrimination [14]. We bridge the gap between classical, analytical methods [15] and Monte Carlo-based approaches [16]. Classical methods may have difficulty managing non-analytical model functions and data-driven Monte Carlo approaches come at a high computational cost. We replace the original, parametric models with probabilistic, non-parametric Gaussian process surrogates learned from model evaluations. The surrogates are flexible regression tools that extend classical analytical results to non-analytical models, while providing us with model prediction confidence bounds and avoiding the computational complexity of Monte-Carlo approaches.

Multi-objective optimization [17, 18]. We make novel extensions to Bayesian multi-objective optimization in the case of one analytical objective function and one black-box, i.e. simulation-based, objective function. The resulting method has been applied to a bone neotissue application [19] and a more general test suite.

Scheduling plant operations under uncertainty. For processes with equipment degradation, we use Gaussian processes to approximate large-scale, mixed-integer optimization problems.

We close by offering a broad outlook on applying probabilistic surrogate models to chemical engineering. Statistical machine learning has recently attracted significant interest in process systems engineering [20]. Here we show that state-of-the-art research in Gaussian processes [21, 22] and probabilistic modeling more generally [23] can have a big impact on chemical engineering.

References

[1] Bhosekar A, Ierapetritou M. Advances in surrogate based modeling, feasibility analysis and and optimization: A

review. *Comput Chem Eng*. 108: 250-267, **2018**.

[2] Caballero JA, Grossmann IE. An algorithm for the use of surrogate models in modular flowsheet optimization. *AIChE J*. 54(10):2633-2650, **2008**.

[3] Lang Y, Zitney SE, Biegler LT. Optimization of IGCC processes with reduced order CFD models. *Comput Chem Eng*. 35(9):1705-1717, **2011**.

[4] Cozad A, Sahinidis NV, Miller DC. Learning surrogate models for simulation-based optim-ization. *AIChE J*. 60(6):2211-2227, **2017**.

[5] Boukouvala F, Ierapetritou M. Surrogate-based optimization of expensive flowsheet modeling for continuous pharmaceutical manufacturing. *J Pharm Innov*. 8: 131, **2013**.

[6] Soepyan FB, Cremaschi S, Sarica C, and others. Estimation of percentiles using the Kriging method for uncertainty propagation. *Comput Chem Eng*. 93:143-59, **2016**.

[7] Shokry A, Ardakani MH, Escudero G, Graells M, Espuña A. Dynamic Kriging-based fault detection and diagnosis approach for nonlinear noisy dynamic processes. *Comput Chem Eng*. 106:758-76, **2017**.

[8] Tran AP, Georgakis C. On the estimation of high-dimensional surrogate models of steady-state of plant-wide processes characteristics. *Comput Chem Eng*. DOI 10.1016/j.comp-chemeng.2018.02.014, **2018**.

[9] Boukouvala F, Hasan MF, Floudas CA. Global optimization of general constrained grey-box models: new method and its application to constrained PDEs for pressure swing adsorption. *J Glob Optim*. 67(1-2): 3-42, **2017**.

[10] Boukouvala F, Floudas CA. ARGONAUT: AlgoRithms for Global Optimization of coNstrAined grey-box compUTational problems. *Optim Lett*. 11(5): 895-913, **2017**.

[11] Pearson RK, Pottmann M. Gray-box identification of block-oriented nonlinear models. *J Process Contr*. 10(4):301-315, **2000**.

[12] Cozad A, Sahinidis NV, Miller DC. A combined first-principles and data-driven approach to model building. *Comput Chem Eng*. 73: 116-27, **2015**.

[13] Boukouvala F, Misener R, Floudas CA. Global optimization advances in mixed-integer nonlinear programming, MINLP, and constrained derivative-free optimization, CDFO. *Eur J Oper Res*. 252: 701-727, **2016**.

[14] Olofsson S, Deisenroth MP, Misener R. Design of Experiments for Model Discrimination Hybridising Analytical and Data-Driven Approaches. *arXiv:1802.04170*. **2018**.

[15] Buzzi-Ferraris G, Forzatti P, Emig G, Hofmann H. Sequential experimental design for model discrimination in the case of multiple responses. *Chem Eng Sci*. 39(1): 81-85, **1984**.

[16] Vanlier J, Tiemann CA, Hilbers PAJ, van Riel NAW. Optimal experiment design for model selection in biochemical networks. *BMC Syst Biol*, 8(20), **2014**.

[17] Olofsson S, Mehrian M, Geris L, Calandra R, Deisenroth MP, Misener R. Bayesian Multi-Objective Optimisation of Neotissue Growth in a Perfusion Bioreactor Set-Up. *Computer Aided Chemical Engineering*. 40: 2155-2160, **2017**.

[18] Beykal B, Boukouvala F, Floudas CA, Pistikopoulos EN. Optimal design of energy systems using constrained grey-box multi-objective optimization. *Comput Chem Eng*. DOI 10.1016/j.compchemeng.2018.02.017, **2018**.

[19] Mehrian M, Guyot Y, Papantoniou I, Olofsson S, Sonnaert M, Misener R, Geris L. Maximizing neotissue growth kinetics in a perfusion bioreactor: An *in silico* strategy using model reduction and Bayesian optimization. *Biotechnol Bioeng*. 115(3):617-29, **2018**.

[20] Lee JH, Shin J, Realff MJ. Machine learning: Overview of the recent progresses and implications for the process systems engineering field. *Comput Chem Eng*. DOI 10.1016/j.compchemeng.2017.10.008, **2017**.

[21] Damianou A, Lawrence N. Deep Gaussian processes. In *Artificial Intelligence and Statistics*. 207-215, **2013**.

[22] Deisenroth MP, Ng JW. Distributed Gaussian processes. In *32nd International Conference on International Conference on Machine Learning*. 37: 1481-1490, **2015**.

[23] Ghahramani Z. Bayesian non-parametrics and the probabilistic approach to modelling. *Phil Trans R Soc A*. 371(1984):20110553, **2013**.

533740 Connecting Experimental Conditions with Chemical Structure

Abstract:

Reaction databases provide a great deal of useful information to assist planning of experiments, but do not provide any interpretation or chemical concepts to accompany this information. In this talk, reactions are labeled with experimental conditions, and network analysis will show that consistencies within clusters of data points can be leveraged to interpret underlying concepts that organize the information. In particular, this analysis will show why particular choices of solvent are effective in enabling specific organic reactions (Friedel-Crafts, aldol addition, Claisen condensation, Diels-Alder, and Wittig), including variations in solvent choice within each reaction class. Our strategy will automatically craft succinct sentences or rules of thumb which describe the observed solvent selection rules. An example of network analysis is shown in Figure 1, where a Friedel-Crafts reaction breaks into clusters which depend on choice of catalyst, substrate, and stability of the reaction for/against protic solvent. These observations, typically belonging to the intuition of a chemist, are derived directly from the network.

Figure 1. Friedel-Crafts reactions network.

530881 Symbolic Regression of Alpha Functions for Cubic Equations of State

Abstract:

Since Van der Waal first proposed a formulation for a thermodynamic cubic equation of state, modifications to the thermodynamic cubic equations of state have focused on the alpha function of the attraction term to model the temperature dependence of attraction for real fluids [1]. The alpha function is dependent on reduced temperature and density and acentric factor. Modifications to Peng Robinson and Redlich-Kwong forms of the cubic equation have focused on the functional form of the alpha function fit prior to fitting the attraction and repulsion values for the cubic equation. Traditionally, cubic equations are derived from the critical point which leads to errors in liquid density and in the critical region. Alternatively, when fit to chemical-specific liquid density and vapor pressure data, the equations fail to represent the critical constraints [2].

To improve the development of cubic equations and their regression, we focus on a data-driven constrained symbolic regression to simultaneously determine the model form of the alpha function and fit parameters of the cubic equation. Symbolic regression learns both the model structure and parameters to model a data set, unlike traditional regression that limits the scope of the regression to a fixed functional form [4]. The regression only requires the specification of a set of operators and operands ($+$, $-$, $*$, $\div$, $\exp(\cdot)$, $\log(\cdot)$, $(\cdot)^2$, $(\cdot)^3$, $\sqrt{\cdot}$, etc.) to flexibly develop new functional forms that accurately represent the data. Symbolic regression has typically been performed with genetic programming [4], but recent developments in applying global deterministic approach have shown improved fitting metrics, such as sum of squared error and other information criteria [3]. We use this approach to apply symbolic regression to pure fluids to determine new alpha function forms for cubic equations of state and compare their accuracy to the data, specifically in liquid density and supercritical regions, and adherence to theoretical thermodynamic properties with other alpha modifications of cubic equations of state.

Reference cited:

[1] Valderrama, J. O. The State of the Cubic Equations of State. *Ind. Eng. Chem. Res.* 2003, 42, 1603-1618.

[2] Twu, C.H. & Sim, W.D. & Tassone, V. (2002). Getting a Handle on Advanced Cubic Equations of State. Chemical Engineering Progress. 98. 58-65.

[3] Cozad, A. & Sahinidis, N.V. (2018). A global MINLP approach to symbolic regression. Mathematical Programming, to appear.

[4] J. R. Koza. Genetic Programming: On the Programming of Computers by Means of Natural Selection. MIT Press, 1992.

538152 Predicting Hydrogen Storage in Half-a-Million MOFs Via Machine Learning

Abstract:

Metal-organic frameworks (MOFs) are promising solid-state adsorbents thanks to their high gravimetric capacities. However, realizing a high volumetric H_2 adsorption capacity, balanced with a high gravimetric density, is one of the main barriers of the successful application of MOFs as solid-state adsorbents. A large database of half-a-million MOFs consisting of real and hypothetical compounds was compiled and screened using semi-empirical and atomistic (grand canonical Monte Carlo) techniques. Several machine learning (ML) algorithms were benchmarked for their ability to predict hydrogen storage in MOFs at multiple conditions. The top performing algorithm, extremely randomized trees, was applied to rapidly identify MOFs with high usable H_2 storage capacities across the entire database. A combinatorial approach was then used to understand the importance of crystallographic properties and training set size. This approach identifies the number and combination of crystallographic features needed to achieve the most accurate predictions. Finally, we assess multilinear regression models for their ability to out-perform the well-known Chahine rule for predicting H_2 uptake.

537160 Inverse Design of Nanoporous Adsorbents for Gas Separation Applications

Abstract:

Significant research efforts have been undertaken for *in silico* screening of nano-porous materials in applications such as methane storage, carbon capture[1-2]. Both theoretical insights and data-driven approaches have been used to develop structure property relationships for properties such as gas storage, selectivity, heat of adsorption[3-6]. Balancing multiple material property tradeoffs is important for both commercial and novel applications. Optimizing a completely algebraic formulation of a process can provide optimal material property targets. While these may not correspond directly to an existing material, possible structures can be predicted inversely by exploiting similarity[7] between structures and properties of adsorbent materials.

In this work, the optimal adsorbent properties (e.g. adsorption isotherms, heat of adsorption etc.) are determined by optimizing a completely discretized non-linear programming model of a pressure swing adsorption process for a desired objective such as cost or productivity subject to process constraints such as purity and recovery. The bounds on the material properties are set based on the isotherm data obtained from Grand Canonical Monte Carlo (GCMC) simulations of structures in the IZA-SC database. This approach is advantageous as all the material alternatives do not have to be individually considered. Moreover, the optimal material properties can be recalculated for changes in process and feed conditions with relative ease. For the reverse estimation of structural properties, a surrogate model based approach[8] is employed. It relates the differences in structural properties such as channel diameter, pore size, accessible surface area and volume between pairs of pure silica zeolites to the difference in their adsorption properties. Based on this, the appropriate structural features corresponding to the target material properties are then obtained. The salient feature of this approach is to design materials for optimal operation of processes such as CO_2 capture, methane storage.

References:

(1) Colon, Y. J.; Snurr, R. Q. High-Throughput Computational Screening of Metal-Organic Frameworks. *Chem. Soc. Rev.*, **2014**, 43 (16), 5735–5749.

(2) Lin, L.-C.; Berger, A. H.; Martin, R. L.; Kim, J.; Swisher, J. A.; Jariwala, K.; Rycroft, C. H.; Bhown, A. S.; Deem, M. W.; Haranczyk, M. and Smit, B. In silico Screening of Carbon-Capture Materials. *Nat. Mater.*, **2012**, 11 (7), 633–641.

(3) Wilmer, C. E.; Farha, O. K.; Bae, Y-S.; Hupp, J. T. and Snurr, R. Q.; Structure–property Relationships of Porous Materials for Carbon Dioxide Separation and Capture, *Energy Environ. Sci.*, **2012**, 5, 9849-9856.

(4) Garcia, E. J.; Perez-Pellitero, J.; Jallut, C. and Pirngruber, G. D. How to Optimize the Electrostatic

interaction between a Solid Adsorbent and CO_2. *J. Phys. Chem. C*, **2014**, 118, 9458-9467.

(5) Bai, P.; Jeon, M. Y.; Ren, L.; Knight, C.; Deem, M. W.; Tsapatsis, M. and Siepmann, J. I. Discovery of Optimal Zeolites for Challenging Separations and Chemical Transformations using Predictive Materials Modeling. *Nat. Commun.*, **2015,** 6, 5912.

(6) Fernandez, M.; Boyd, P.G.; Daff, T. D.; Aghaji, M. Z. and Woo, T. K. Rapid and Accurate Machine Learning Recognition of High Performing Metal-Organic Frameworks for CO_2 Capture. *J. Phys. Chem. Lett.* **2014**, 5, 3056-3060.

(7) Martin, R. L.; Willems, T. F.; Lin, L.-C.; Kim, J.; Swisher, J. A.; Smit, B.; Haranczyk, M. Similarity-Driven Discovery of Zeolite Materials for Adsorption-Based Separations. *ChemPhysChem*, **2012**, 13, 3595–3597.

(8) Bajaj, I.; Iyer, S. S. and Hasan, M. M. F. A Trust Region-based Two Phase Algorithm for Constrained Black-box and Grey-box Optimization with Infeasible Initial Point. *Comput. Chem. Eng.*, **2017**, doi: 10.1016/j.compchemeng.2017.12.011.

530836 Classifying Antimicrobial and Multifunctional Peptides with Machine Learning

Abstract:

Antimicrobial and antifouling peptides are highly desirable biomolecules, but they are relatively rare among all known peptides. It is not intuitively apparent from an amino acid sequence whether a given peptide will exhibit these properties. Thus, it is of interest to develop a data-driven quantitative structure-activity (QSAR) model to predict potentially antimicrobial and antifouling peptides. Machine learning techniques are well-suited to such a task, and have the advantage of a robust, established pool of algorithms for learning. Here, we exhibit Bayesian network QSAR models that allow us to accurately predict whether an input peptide may exhibit antimicrobial properties, given its sequence.

529531 A Systematic Procedure for Designing Training Data for Molecular Property Prediction

Abstract:

Organic material design requires thoroughly exploring chemical compound space to gain desired property information. The astronomical size of molecule space, however, makes it impossible to use experiments of quantum chemistry to evaluate every molecule in the space; consequently, data-driven semi-empirical models are required to calculate molecular properties rapidly. To this end, the focus of this talk is two-fold. First, we show that the group contributions approach can be generalized using a combination of Cheminformatics-based path fingerprints and sparse modeling techniques to derive an ab initio data-driven model to accurately estimate heats of atomization of small organic molecules, reaches the mean absolute error of 1.59 kcal/mol and 2.61 kcal/mol for QM7 and QM9 dataset excluding molecules with fused rings. Further, we show that modern experimental design tools and cheminformatics-based subset selection techniques can be combined to systematically minimize the amount of data needed to train a model. We specifically show that, given a set of molecules for which data-driven models are sought, a carefully chosen subset of even 2- 10% of original dataset is sufficient to train a model that is as accurate as one that is trained on >80% of the data.

528153 Designing Proteins with Enhanced Antifreeze Activity Using SimulatedÂ Directed Evolution

Abstract:

Found in a variety of organisms, from cold water fish to arctic beetles, antifreeze proteins (AFPs) are a diverse class of proteins that depress the non-equilibrium freezing point of water. AFPs have been of scientific interest for over 30 years due to their role in organism survival at low temperatures, and have potential applications in cryopreservation, cryosurgery, clathrate formation inhibition, and specialty coatings. Despite a large body of experimental and theoretical literature accumulated on AFPs, there exists no widely applicable method for accurately predicting AFP activity - making intelligent design of new antifreeze materials challenging. To remedy this, we perform molecular dynamics simulation with a collection of different types of AFPs (ranging from ~30 to 300 residues in length) in liquid water just below the freezing point. By analyzing the dynamics of hydrogen bonds in water near the protein in conjunction with the geometry of the protein structure, we establish a general and quantitatively accurate correlation between simulation and experimentally measured antifreeze activity. Implementing this correlation in a directed evolution algorithm, we identify several key mutations that exhibit more than five time the predicted activity than the wild type protein.

538791 Transition State Geometry Prediction Using Neural Embeddings of Transition State Graphs

Abstract:

Transition State Theory (TST) with modern quantum chemistry methods allow reaction rates coefficients to be calculated with high accuracy, but finding the transition state is computationally expensive and difficult to automate. The key step in TST is to determine the transition state (TS) geometry, which is mostly determined by reaction family, reaction center, and reactant geometries. Inspired by neural word embeddings techniques, we introduce rxn2vec, which is an unsupervised machine learning approach to create vector representations of transition states. We build upon a neural embedding framework named graph2vec [1] to learn data-driven distributed representations of arbitrarily sized graphs. In this case, fully optimized TS geometries generated by our automated TST calculator (AutoTST) [2] are converted to undirected graphs using connectivity matrices. These training data then provide the framework to predict key TS reaction center geometries. We demonstrate our tool on a set of reactions, predict key distances from these reaction, and implement AutoTST to perform partial geometry optimizations to converge on a finalized geometry. So far this tool can be used to predict the reaction centers of three reaction families, namely hydrogen abstraction, intramolecular hydrogen migration, and radical addition to multiple bond.

[1] graph2vec: Learning Distributed Representations of Graphs
A. Narayanan, M. Chandramohan, R. Venkatesan, L. Chen, Y. Liu, S. Jaiswal (2017)
arXiv:1707.05005 [cs.AI]

[2] Automated Transition State Theory Calculations for High-Throughput Kinetics
Pierre L. Bhoorasingh, Belinda L. Slakman, Fariba Seyedzadeh Khanshan, Jason Y. Cain, and Richard H. West. J. Phys. Chem. A.121 (37), 6896-6904, (2017) https://doi.org/10.1021/acs.jpca.7b07361

538578 Transforming Undergraduate Chemical Engineering Education - a Report from NSF RED (Revolutionizing Engineering & Computer Science Departments) Projects

Abstract:

While industry has long called for changes to how engineers and computer scientists are prepared, much of the core coursework remains virtually unchanged. To address this mismatch between industry and education, NSF called for proposals that aim to ***RE****volutionize Engineering and Computer Science* ***D****epartments*, serving as leaders in institutional change. This presentation will provide short reports from three chemical engineering departments that received grants in the NSF RED competition in the 2015, 2016 and 2017 cohorts. The talks will present various models of structural change at the departmental level, focused on broadening participation, improving student learning, and reorienting faculty beliefs and practices.

538605 Ready, Set, Speak! Classroom Activities for Developing Oral Communication Skills

Abstract:

Oral communication is an essential professional skill and while many classes and programs emphasize oral communication practice through regular use of formal presentation assignments, dedicated time to practicing specific oral communication skills is often not included. In a dedicated course on technical and professional communication, a series of impromptu speech and role playing activities are used to both draw studentsâ€™ attention to specific oral communication skills and provide opportunities for dedicated practice of those skills. Each activity has a specified set of constraints, and often uses extreme scenarios (e.g., having only 5 seconds to review a slide you have never seen) to allow students to assess their own tendencies, as well as develop a â€œfeel forâ€ how they can most successfully structure their informal and formal oral communication. Areas of focus for the activities include eliminating use of filler words, maintaining audience engagement through eye contact, avoiding look at slides excessively, as well as use of body language or visual representations of information. In each activity, students are able to practice their skills, and discover the underlying reasons behind why certain oral communication tendencies occur (e.g., why and when we tend to use filler words) and develop strategies for improvement. Year after year, these activities continue to be student favorites, often described as one of the most immediately impactful experiences in the course.

536839 Know Your Audience: Expanding the Range of Scientific Communication in the Classroom

Abstract:

While the development of communication skills is critical to a successful career in engineering, teaching these skills continues to be one of the biggest challenges to engineering faculty. In a traditional chemical engineering curriculum, students focus on communication of results through formal lab reports and presentations that are directed at expert audiences (i.e., other students and faculty). While these assignments can successfully represent the technical understanding of the student, they are not fully representative of the communication skills required of a bachelor level engineer in industry. The critical difference between what is taught in the classroom and what is required in an industry career is the audience. To better prepare students for their careers, it is crucial for students to learn to communicate to both scientific and non-scientific audiences. As a result, assignments were developed to allow students to practice communication of technical content to expert, business and non-technical audiences. In each of these assignments, students are challenged to understand how their audience can change not only the language that must be used, but also how the work should be motivated to interest the reader. Through student-led peer-revision sessions, students reflect on the effectiveness of an assignment at targeting a specific audience. These informal discussions allow students to solidify their understanding of the material and their communication approach, resulting in improvement of the assignment prior to submission for grading. While it is impossible to prepare students for every form of communication that will be required of them in industry, expanding the range of communication assignments in the classroom helps to reinforce a key requirement for successful communication: knowing your audience.

534442 The Art of Communication: Ways to Incorporate Active Learning to Develop Student Communication

Abstract:

Communication is a skill that students are expected to have upon graduation, however there are usually minimal opportunities for students to practice delivering oral presentations and writing technical reports. These skills are often assessed at the culmination of capstone projects and the conclusion of an experiment for unit operations, with minimal preparation. In this presentation there will be discussion of think-pair-share and a jigsaw activity that help students with creating effective PowerPoint presentations and improve student technical writing skills.

536092 The Technical Memo

Abstract:

Our Material and Energy Balances II course has been structured to meet part of the composition requirement for general education at our university. Our BS CHE program has identified 5 key characteristics of effective written technical communication that should serve graduates well in industry and in graduate or professional school.

1. Follows format conventions for figures, graphs, tables, and values.
2. Organized to convey key information first, followed by supporting detail.
3. Uses terminology exactingly and writes in a concise, direct style.
4. Goes beyond a display of data / information to provide analysis.
5. Cites and uses literature to provide supporting data or to support analysis.

For six years, we have used the "technical memo" in Material and Energy Balances II as a vehicle to develop these technical communication skills in our students, involving four faculty members in the writing instruction over 12 offerings of the course. Our emphasis in the course is on organization to convey key information first, then follow up with supporting detail. From the students' work on these assignments, we have gained insight into their perceptions of what counts as "key information" and into some of their misconceptions about engineering concepts as they attempt to provide supporting detail. We have learned to better define what we mean by good writing, and about the need to focus our attention on specific aspects of writing to see the students develop them successfully. We continue to debate best approaches.

In this presentation, we will share assignments, rubrics, aggregated assessment and evaluation, and selected (and anonymous) student work. We will also describe how this course fits into our overall development of technical communication skill within the curriculum, and some of the debates among instructors as we have found ourselves divided by our common purpose.

535540 Starting Early and Small with Technical Communication: Sophomores, Plots, and Captions

Abstract:

One obstacle to development of concise technical communication skills is the years of aiming for quantity of words at the expense of quality of content. In this work, that conditioning is challenged by focusing on a very small writing assignment â€“ a caption that encapsulates all the essential information to describe a plot, without any extraneous information. The assignments are performed by students in a material and energy balances course, and the vehicle for the caption-writing is a series of assignments involving spreadsheet calculations for which the solutions are to be submitted as plots. In addition to teaching succinct communication, this practice provides an early opportunity to teach the basic mechanics of technical writing and to correct common errors before they become ingrained. Instruction on proper formatting and presentation of graphs is included. Critical thinking skills applied to communication are developed as students distinguish between essential and extraneous information. Finally, the inherent shortness of the assignment facilitates efficient grading and quick feedback, which can even be done in class by peers.

537015 Technical Communications Emphasis in Senior Chemical Engineering Laboratory Course

Abstract:

Recent changes in our departmentâ€™s chemical engineering curriculum have provided a natural opportunity to investigate the question: Can technical communication instruction be successfully integrated into an existing technical course?

For several years, a technical communications course was a required prerequisite to the senior laboratory course in chemical engineering; the technical communications course was recently cut from the curriculum due to budget pressures, providing an incentive to integrate technical communication instruction into the laboratory course. Many, if not most, institutions experience similar constraints that make it impractical to offer a dedicated technical communications course. By necessity, technical courses that have a strong communication emphasis prioritize the instruction and mastery of technical content, whereas dedicated courses offer an opportunity to focus deeply on communication content.

We evaluate the process, now underway for a third semester, of integrating technical communication instruction into the capstone laboratory course. Our first steps have been 1) to create detailed student rubrics for written and oral presentations that reflect key attributes of successful communications and 2) to bring technical communication content to weekly interactive lectures, with attention given to expectations on the course deliverables: written and oral lab reports. Some topics we have covered include: consideration of audience, purpose, and context, technical communication as logic, writing with concision, creating effective slides, and presenting with consideration to voice modulation and body language. In addition, our lectures have covered professionalism, safety, and ethics.

536645 Teaching Technical Communication to Chemical Engineering Undergraduate Students through the Use of a Learning Community and Linked Courses

Abstract:

How can technical communication principles be integrated into the chemical engineering undergraduate curriculum? For almost seven years, a learning community has linked the first of two Unit Operations laboratory courses to a technical communication course. In this learning community, students are concurrently enrolled in the courses, and the instructors work together to make sure that common elements are required in assignments. As part of the learning community, peer mentors guide students through various details about the types of reports required in the courses, as well as give feedback to report drafts.

In a previous paper [1], after two years of experience with the learning community, it was reported that significantly higher student scores were earned on written reports by students who were concurrently enrolled compared to those who were not. In the current work, four additional years of experience with the learning community are investigated and new questions are addressed to evaluate the impact of this program. Recent updates to the technical communication course include an emphasis on visual communication in both written and oral reports, and a change of report evaluation techniques in the technical communication course. The evaluations are now conducted in a "face-to-face" format. In this format, students spend up to three hours over the course of the semester in individual meetings with the English instructor, receiving feedback during review and grading of the submitted reports.

In this talk, we intend to answer the following questions: are the trends observed early in the program continuing as before? Do the students from the linked sections show retention of the information as they progress into later chemical engineering courses, such as the second Unit Operations laboratory and Process Design? How has the new report assessment process impacted student learning and perception of the importance of being able to communicate effectively?

[1] Loveland, S.D., S.D. McGough, "A Learning Community: Teaching Technical Writing in the Chemical Engineering Unit Operations Laboratory," Proceedings of the 2014 ASEE North Midwest Section Conference, October 16-17, 2014, Iowa City, IA.

538290 Adapting the COPE (Clarity, Organization, Precision, and Economy) Framework As a Writing Guide and Assessment Framework

Abstract:

The COPE (Clarity, Organization, Precision, and Economy) framework by David Adams is a fourteen point checklist for effective writing. Initially tested within the communications workshops for a Research Experience for Undergraduates (REU) site, the COPE framework has been adapted as both an assessment tool and a manual to provide pointed feedback to student writing. A summary of the fourteen points will be presented along with examples of feedback and assessment procedures.

543565 Lessons Learned during the Tenure-Track Process

Abstract:

The talk will focus on the lessons I learned over the years on how to succeed as a junior faculty. I will discuss strategies for effective teaching, successful recruitment and mentoring of graduate students, as well as winning research grants including the NSF CAREER.

542553 Ten Hints for Better Teaching

Abstract:

Everyone who wants to can become a good teacher. Although it takes a few years of practice to become a good teacher, new teachers can start using many of the practices of good teachers immediately. This talk will outline ten methods that can be employed immediately to improve teaching. There will be time to answer questions about teaching.

540912 Education and Accreditation Committee Discusses Updates and Insights Regarding ABET Accreditation

Abstract:

Education and Accreditation Committee members will discuss recent updates and insights regarding ABET accreditation. Topics include: program preparation, program evaluator preparation, the visit schedule, post-visit interactions, common criterion issues during a visit, recent changes to accreditation, and accreditation resources. An important aspect this year is a presentation on the new student outcome changes beginning in the 2019-20 accreditation cycle. This session is particularly beneficial for faculty members of programs undergoing ABET accreditation during the next several years and for those interested in becoming program evaluators.

537752 Client-Sponsored and Team-Defined Senior Capstone Projects

Abstract:

The Departmentâ€™s capstone experience allows students to apply their chemical engineering education to solve an open-ended design problem. In the past, teams have applied the engineering design process to paper-based projects, such as the AICHE-sponsored design challenge. In recent years, we have collaborated with local industrial sponsors seeking to optimize part of a manufacturing process. Students must coordinate site visits with the client to observe the process under review and work closely with the industrial partner to deliver expected results. Additionally, student teams are required to perform hands-on experimental trials in the Unit Operations laboratory to test specific design hypotheses. Recent project examples include: 1) Assess efficiency/improvement opportunities for manual assembly line at the B.Braun, Dominican Republic site. 2) Provide a preliminary process design for the recovery of edible protein from cheese whey at Klein Farms. 3) Assess efficiency/improvement opportunities for the watercolor assembly line at Crayola. 4) Assess efficiency/improvement opportunities for bottle filling line at FMI Cosmetics.

This year, the Department expanded the capstone project offerings to include a Team-Defined project. These student-proposed design projects are comparable in scope to client-sponsored projects and can be extensions of student summer internships. Team-Defined projects also present a great opportunity to develop an entrepreneurial idea of interest to members of the team. In the case of an entrepreneurial idea, there is an added requirement of a more detailed pro forma for the proposed business. Team-defined project examples include: 1) Provide a preliminary process design for a commercial plant for 1) kombucha production, 2) apple cider vinegar production, and 3) orange juice concentrate production. Student teams are required to perform hands-on experimental trials in the Unit Operations laboratory to test specific design hypotheses, and also simulate the commercial scale plant in ASPEN.

In this presentation, several projects will be highlighted to demonstrate achievement of course objectives, which are: presenting technical material to an audience at an appropriate level of difficulty and approach; working in teams to effectively and ethically overcome engineering problems and technical challenges; demonstrating project management, organization, and documentation skills for the effective operation of an engineering development team; performing an economically sensitive process or product design project to demonstrate an understanding of cost, profit, regulation, and market issues; and preparing an exhaustive engineering technical report that relates a project in such a manner as to satisfy the goals of clarity and completeness in detail and presentation.

In our experience, client-sponsored and team-defined projects have been well received by students looking for a practical application of their education in their final capstone experience, and present an acceptable alternative to paper-based design.

524225 Embedding Engineering Research in Process Design Education

Abstract:

Technical literature search is undoubtedly one of the most important components of chemical process design. However, many undergraduate students graduate without knowing how to effectively search, synthesize, and utilize knowledge from vetted technical literature (e.g journals, patents, standards, handbooks) in engineering design settings. Students tend to over rely on surface-level Google searches, while underutlizing powerful information discovery toolsets that are available to them through organizational subscriptions, such as Knovel and Engineering Village. Research skillsets are most successfully learned in context of an actual project, rather than a one-off lecture. To this aim, Research Sprints are a fast-paced student-centered learning experience that combine elements of design thinking, agile processes, and engineering research together, to layer in technical acumen with collaboration and communication skillsets.

Knovel is an engineering decision support tool that builds upon a rich corpus of engineering textbook and material property content to serve engineering analytics use cases. Engineering Village complements Knovel with deep indexing of engineering journal, conference, IP, standards, and funding literature. In this talk, we will demonstrate how Knovel & Engineering Village can be used together into the classroom to expose students to technical literature and apply the insights in engineering design.

533875 Effective PBL in Senior Design: Examples & Lessons

Abstract:

With problem-based learning (PBL) students learn by solving an open-ended problem. Education research suggests that PBL is one of the trickiest instruction modes to implement. For example, some studies found that PBL success is more sensitive, compared to other instruction methods, to the subject matter 'expert level' of the instructor. PBL furthermore tends to fail more drastically when it becomes overly student self-directed (Prince, Journal of Engineering Education, 93(3), 2004). Virtually every BS Chemical Engineering program in the USA culminates with a capstone design module that is inherently a PBL exercise. Is this experience always, or even typically, as effective a learning experience as it could be?

This presentation will share a few examples drawn from two new BS Chemical Engineering programs, the first at the Rochester Institute of Technology (first graduating class 2013) and the second from the University of Iceland (first class 2017). New programs afford a unique 'blank sheet' opportunity to assemble a complete curriculum from scratch with no historical baggage and test alternative approaches.

The presentation will include a brief survey of design sequence curriculum choices across institutions in the USA. Selected examples of tried-and-tested design problems with specific learning outcomes will be shared. The importance of curriculum integration will be illustrated.

537284 Incorporating Laboratory Pilot Plant Data and Customized Software Simulations into Chemical Engineering Design Education at Vanderbilt

Abstract:

Russell F. Dunn, Ph.D., P.E.

Professor of the Practice

Department of Chemical and Biomolecular Engineering

Vanderbilt University

Nashville, TN 37235-1604

Ph: 615-343-4268

russell.dunn@vanderbilt.edu

Matt J. Lang, Ph.D.

Professor

Department of Chemical and Biomolecular Engineering

Vanderbilt University

Nashville, TN 37235-1604

Ph: 615-345-7493

matt.lang@vanderbilt.edu

Bryan R. Beyer

Adjunct Instructor

Department of Chemical and Biomolecular Engineering

Vanderbilt University

Nashville, TN 37235-1604

Ph: 615-343-4268

bryan.r.beyer@vanderbilt.edu

Incorporating Laboratory Pilot Plant Data and Customized Software Simulations into Chemical Engineering Design Education at Vanderbilt

Integrating unit operations laboratories with scale-up design assignments offers unique perspectives for the students to connect equipment pilot equipment operation to industrial scale design. Two-week laboratory modules involving reverse osmosis and distillation technology, respectively, are currently run by undergraduates students at Vanderbilt as part of their senior unit operations laboratory course. Students then use this pilot scale data to generate a traditional laboratory report covering these respective technologies. An additional week is then spent by the students using customized water network design software is coupled with the reverse osmosis technology to design an industrial scale water interception and recycle network. Likewise, customized heat exchange network software and Aspen simulations are coupled with the distillation technology to design an industrial scale distillation column with heat integration. Two examples illustrating these approaches will be shown.

534970 Implementation of Joint Senior and First-Year Design Projects

Abstract:

Students in their final year of chemical engineering complete a design project that requires them to use the material learned over the course of their studies. In their undergraduate curriculum, the students acquire design and project management experience but very little leadership experience. At the same time, incoming first-year chemical engineering students have a difficult time connecting their introductory mathematics and science courses to their future engineering courses. While freshman students take an introduction to chemical engineering course with a design project, they lack the technical knowledge and skills to do meaningful projects. To help with all of these problems, we created a joint design project where first year students work as interns on a senior design team project that focuses on product design. The two classes share a one hour meeting time each week, as well as having separate class time. The senior students act as supervisors and mentors for the first-year students and give assignments and provide feedback for the first-year students. The design product is divided into four phases: (1) product idea development (2) final product selection (3) process design, and (4) final design. In addition to the assignments from the senior students (emphasis in phase 1 and 4), first-year students receive overarching instructor given assignments within each phase of the project. For phase 1, both groups conduct market research to determine product opportunity gaps. The first-year students are required to submit a report detailing their research and market analysis, but the specific research assignments were given by the mentor teams. Additionally, the first-year reports were submitted to the senior teams for feedback prior to being submitted to the instructor. For phase 4, each member of the first-year team becomes an expert on one of the following areas: safety, economic impact, or environmental impact. The student then researches and reports on this topic to both the instructor and the mentor teams. In phase 2, the first-year teams work with the mentor teams to create a weighted selection matrix and narrow down product ideas, culminating in a sales pitch, advocating for one particular product selection, to the mentor teams. Mentor teams select the final design idea. Phase 3 will have the most separation between the teams with the senior teams creating the actual product process design, and the first-year teams creating a product in a hands-on laboratory, a simplified block diagram, and a material balance for the process. In the first two years, both senior and first-year teams were surveyed at the end of the semester. This paper focuses on the results of implementation of the project in its first two years, and changes made for the second year implementation based on feedback from the first year.

530248 Core Chemical Engineering Graduate Bridging Course: Development and Early Evaluation

Abstract:

This work discusses the development of an online chemical engineering bridging course, followed by an early evaluation of the first offered semester in Fall 2018. The two-semester online course covers main chemical engineering topics including material and energy balances, thermodynamics, transport, and reactor design. The primary objective of the course is to prepare students with backgrounds in chemistry, biology and other non-chemical engineering fields for graduate study in chemical engineering. Engineers and scientists from industry who want comprehension of core concepts of chemical engineering and are looking for a continuing education option will also benefit from this course. The delivery of the course incorporates short course modules utilizing multimedia features (e.g. text and short videos) along with modern pedagogical and assessment technologies (e.g. screencasts, light board recordings, and conceptual quizzes with instant automated feedback). Individual live recorded chatrooms are linked to each topic in the course which allow students and the instructor to collaborate in real time and discuss the course material in more depth.

Preliminary background and development details for this course were presented as a Work-in-Progress at the 2018 ASEE annual meeting in Salt Lake City, UT, and feedback and suggestions from the ChE education community were incorporated to the first course offering. Studentsâ€™ ability to achieve the specified course leaning objectives and their overall satisfaction with the teaching methods are considered very important measures to determine course success. Early data-driven analysis and evaluation of the course will be presented. The data collected include mid-term course evaluations, student surveys, academic records, and studentsâ€™ grades on homework and exams. The collected data are categorized based on demographics, course experience and engagement, academic performance, learning perceptions, and motivation. The evaluation is also focused on course structure, student experience, and content for future improvement.

527206 Chemical Engineering Teaching Practicum Course: Reflections after Five Years

Abstract:

We introduced a teaching practicum course for first-year chemical engineering graduate students in 2013. The learning goals for this course are to prepare new graduate students for their upcoming assignments as teaching assistants and to develop communication skills that will be transferrable to their research training. In this talk we will provide a synopsis of how the course has evolved in five years and share our key actions for success, which include using a co-teaching model for the course and rotating instructor assignments among faculty in the department. We have collected outcomes data using a survey instrument for graduate teaching assistants developed by Sohoni, Cho, and French [1]. The survey was administered during the practicum course and midway through the studentsâ€™ first semester of teaching. We find that students show gains in their confidence to effectively communicate with undergraduate students. We will provide additional analysis of student outcomes measured in this survey to describe the impact that the course has made on graduate student education and the undergraduate teaching mission in our department.
[1] Sohoni S, Cho Y, & French DP, A Survey to Capture Needs Assessment for Graduate Teaching Assistant Training, Advances in Engineering Education, Winter 2013, 1 - 20.

534906 Teaching the Data Management Plan (DMP) to Graduate Students

Abstract:

The number and types of recently developed courses on Research Data Management (RDM) demonstrate the need for effective education of graduate students in this area. One topic that is common to most RDM courses is the Data Management Plan (DMP). A DMP commonly means one of two types of documents: 1) the two page DMP that is often required by funding agencies as part of a proposal, or 2) the larger DMP documentation for day-to-day use on a research project. The two DMPs' are designed to address the following major topics: 1) Data that will be created, 2) Data documentation and organization, 3) Data storage and security, 4) Data management after project completion, and 5) Data accessibility for reuse and sharing. The graduate RDM courses generally include a student assignment or term project to prepare one or the other type of DMP, often based on the student's research project. RDM courses offered through libraries often cover the funding DMP while department offered courses typically cover the project DMP. However, there is not a clear consensus in the literature over exactly what the DMP is and how it is used, and no course to date appears to cover both types of DMPs. Obviously this misunderstanding leads to several different approaches for teaching the DMP while also potentially leaving a hole in the student's education. These very different types of DMPs then lead to the question: what is the difference in these DMPs, what type of DMP is being taught in the literature courses, and what should be covered in the DMP topic/assignment of a RDM course?

A RDM graduate course was developed and co-taught by a librarian and faculty member with an active research program.[1] This course was designed to provide the students a focused application of RDM to active research projects. The course included tools for developing both types of DMPs and student assignments on both types of DMPs. Following instruction on the use of DMPtool.com[2], an individual student assignment required preparation of a funding agency type DMP based on the agency providing the funding for the student's project. The Purdue Data Curation Profiles toolkit [3] was then used as a tool to develop a project based DMP for an active research project as an end-of-semester team project.

The focus of this presentation will be an examination of DMPs and how and what should be taught in a graduate RDM course. Examples of DMP instruction from courses reported in the literature will be compared and contrasted. The differences between the two types of DMPs as well as the reasons behind them will be delineated and discussed. Terminology to differentiate the DMPs will be presented. The approaches and outcomes of teaching the two types of DMPs in our course will also be described and compared with the literature. Results and lessons learned from this approach will be discussed.

1. Schmidt, L.O. and J.H. Holles, *A Graduate Class in Research Data Management.* Chemical Engineering Education, 2018. **52**(1): p. 52-59.
2. University of California Curation Center, *Data Management Plan Tool.* 2014; Available from: https://dmptool.org/.
3. Carlson, J., *The Data Curation Profiles Toolkit: User Guide*, in *Data Curation Profiles Toolkit* 2010; Available from: http://datacurationprofiles.org/.

530600 Chemical Accounting with an Open-Access Life Cycle Inventory for Graduate Researchers

Abstract:

Budding researchers are challenged with the daunting task of understanding and analyzing their work in the broad context of the literature. Life cycle inventories (LCIs) serve to categorize resources, map their flow, and evaluate products and processes. These interdisciplinary tools aggregate data to analyze environmental and socioeconomic impacts. They may be used to compare research, highlight areas for improvement and identify potential obstacles. Used widely in industry, many techniques leverage proprietary data and software that acts as a black box. Open-access materials allow for LCIs to be made widely available to students and they may be edited and customized to suit distinct research questions. The analyses can address topics including toxicity, global warming potential, cost, atom economy, and change between iterations. Piloted in a graduate seminar and with summer research students, an excel based LCI proved useful in helping students organize and better understand their work, compare it to the literature, and help suggest next steps. As they are functionally similar to the pre-lab found in undergraduate classrooms, LCIs are important systems to compartmentalize and contextualize data to help students become scholars.

523289 Leadership Education for Engineers

Abstract:

Nearly all engineers work in teams where a leader oversees and guides process/product research and development or manufacturing operations. Exceptional performance in leadership or management positions requires detailed knowledge of the fundamentals of the specific field and related areas, and the ability to apply that knowledge to solve problems. However, these are necessary but not sufficient conditions for career success. When technically-trained individuals undertake a leadership role, frustration is a frequent outcome, despite technical competency and good intentions. This talk will discuss reasons why engineers and scientists often find adaptation into leadership roles disconcerting. Topics and approaches that build awareness in undergraduate and graduate engineers of the requirements and demands in technical leadership positions will be described. Specifically, course and/or workshop content designed to present this information in an interactive, discussion-based approach that addresses situations normally encountered by early and mid career engineers will be discussed.

530168 How We Teach: Chemical Engineering Thermodynamics

Abstract:

The authors present the results of the 2018 AIChE Education Division survey on how chemical engineering thermodynamics courses are taught. The survey contacted all chemical engineering department chairs in the US and Canada and requested that the link be forwarded to faculty who recently taught in the thermodynamics course/course sequence. The report consists of two parts: the statistical and demographic characterization of the course and its content; and the remainder seeks to bring out interesting and effective teaching practices described by survey respondents. Additionally, a limited historical comparison is made between the selected survey results and surveys on the same course conducted in previous years.

530968 Perspectives on Thermodynamics Instruction

Abstract:

Fundamentals of Chemical Engineering Thermodynamics authors, Kevin Dahm and Donald Visco, will share their thoughts and observations on the state of thermodynamics instruction and best practices in cultivating student understanding.

AUTHOR INDEX

AUTHOR INDEX

AUTHOR INDEX